Functional Polymers

INDUSTRY–UNIVERSITY COOPERATIVE CHEMISTRY PROGRAM SYMPOSIA

Published by Texas A&M University Press

ORGANOMETALLIC COMPOUNDS
Edited by Bernard L. Shapiro

HETEROGENEOUS CATALYSIS
Edited by Bernard L. Shapiro

NEW DIRECTIONS IN CHEMICAL ANALYSIS
Edited by Bernard L. Shapiro

DESIGN OF NEW MATERIALS
Edited by D. L. Cocke and A. Clearfield

FUNCTIONAL POLYMERS
Edited by David E. Bergbreiter and Charles R. Martin

OXYGEN COMPLEXES AND OXYGEN ACTIVATION BY TRANSITION METALS
Edited by Arthur E. Martell and Donald T. Sawyer

Functional Polymers

Edited by
David E. Bergbreiter and
Charles R. Martin

Texas A&M University
College Station, Texas

Plenum Press • New York and London

Library of Congress Cataloging in Publication Data

Functional polymers / edited by David E. Bergbreiter and Charles R. Martin.
   p.   cm.
   Proceedings of the Sixth Annual IUCCP Symposium on Functional Polymers, held
March 22–24, 1988, in College Station, Texas.
   Includes bibliographical references and index.
   ISBN-13: 978-1-4612-8096-5   e-ISBN-13: 978-1-4613-0815-7
   DOI: 10.1007/978-1-4613-0815-7

   1. Polymers and polymerization—Congresses. I. Bergbreiter, David E. II. Martin,
Charles R. III. IUCCP Symposium on Functional Polymers (6th: 1988: College Station,
Tex.)
TP1081.F86   1989   89-3932
668.9—dc20   CIP

Proceedings of the Sixth Annual IUCCP Symposium on
Functional Polymers, held March 22–24, 1988,
in College Station, Texas

© 1989 Plenum Press, New York
Softcover reprint of the hardcover 1st edition 1989

A Division of Plenum Publishing Corporation
233 Spring Street, New York, N.Y. 10013

PREFACE

This monograph contains manuscripts, poster abstracts and summary statements representing the contributions of a group of scientists who participated in the sixth annual Texas A&M Industry-University Cooperative Chemistry Program (IUCCP) at Texas A&M University in College Station, Texas, March 22-24, 1988. This symposium on "Functional Polymers" was organized by a university-industrial steering committee consisting of Dr. D. Keene, Hoechst Celanese; Dr. D. E. McLemore, Dow Chemical Company; Dr. B. Frushour, Monsanto Company; Dr. S. Corley, Shell Development; Dr. F. Hoffstadt, BPAmerica; Dr. D. E. Bergbreiter, Texas A&M University; Dr. C. A. J. Hoeve, Texas A&M University; Dr. C. R. Martin, Texas A&M University; Dr. A. Clearfield, Texas A&M University; and Dr. A. E. Martell, Texas A&M University. The symposium itself was generously supported by the industrial companies participating in the IUCCP program. These sponsoring chemical companies include; Shell Development Company, Dow Chemical Company, BPAmerica, Monsato Company and Hoechst Celanese.

The choice of "Functional Polymers" as the subject for this symposium reflects the rapid developments occurring in the broad field of polymer science and the potential for using polymeric derivatives in many new exciting and potentially profitable applications. The invited papers and submitted posters reflect the diversity of this field and include many different topics ranging from biomedical applications of polymers to conducting polymers to use of polymers as lithographic masks and recording media. General topics included in the symposium were: photoresponsive polymers, polymer blends, electronically conductive polymers, polymers catalysts, biomedical polymers and membrane transport and permeability.

The symposium included seventeen invited plenary lectures. Professor R. Macfarlane of Texas A&M University generously agreed to act as a last minute substitute for one speaker who was unable to attend. A highlight of the conference was an evening lecture by Professor Vogl of the Polytechnic Institute of New York on the subject of "Functional Polymers". This illuminating lecture served as a focal point for the whole range of papers presented during this three day meeting.

We thank Mary Martell and Cynthia McBride for assistance with the logistics of the symposium and our students and colleagues with their assistance during the organization of the conference. Charles R. Martin thanks the Office of Naval Research for support during the preparation of this volume.

David E. Bergbreiter
Charles R. Martin

CONTENTS

Poster Abstracts

CONTROL OF PHASE STRUCTURE IN POLYMER BLENDS

D. R. Paul

Department of Chemical Engineering
University of Texas
Austin, TX 78712

INTRODUCTION

Development of new molecules and chemical modifications of existing ones have been the most common ways new challenges for polymeric materials have been approached in the past. These routes have become increasingly complex and expensive over the years, and, thus, alternate ones have become more interesting and attractive. Polymer blending is one such approach that is pressently in a state of rapid scientific and commercial development, as attested by recent books, reviews, and symposia[1-28]. Another related physical approach is polymer-polymer composites in the form of fibers, film, and sheet materials often made via co-extrusion technology[15,29]

Polymer blends can be divided into two major classes based on their thermodynamic phase behavior. Miscible blends are homogeneous to the polymer segmental level and a major factor is the energetics of segmental interaction. Very early in polymer science, it was concluded that miscibility among polymer-polymer pairs was a rare exception[30]. However, research over the last decade has proven this rule to be somewhat overstated[13,14]. It is generally agreed that the thermodynamic basis for miscibility in polymer blends is an exothermic heat of mixing[14-17] since entropic contributions are so small in such systems. A major thrust of research for more than a decade has been to discover new miscible polymer blends and to understand, in terms of component molecular

1

structure, the origin of their energetic interactions so that phase behavior in technologically important multicomponent polymer systems may be better controlled or predicted. Very significant advances have been made in this area. Immiscible blends, on the other hand, have phase domains that consist, in the limiting case, of the corresponding pure components. Major variables affecting properties in these cases are the phase morphology and the degree of adhesion between phases. For polymer-polymer composites the phase morphology is predetermined by the process of fabrication (e.g., parallel layers); however, adhesion between these phases is also a critical factor for successful performance. Very often the adhesion at the polymer-polymer interface is poor; and as a result, blends and composites formed from immiscible polymeric components have inferior macroscopic properties owing to the weakness at this interface. This fact limits the use of such multiphase polymer systems.

The following section outlines some of the incentives for polymer blending and gives examples of the types of problems that can be solved by this route. In every case, the property relationships for blends depend critically on control of phase structture. Therefore, subsequent sections will deal with some of the important scientific issues in this area. The first of these is the selection or design of components when a homogeneous or miscible mixture is needed. The second involves phase separated systems where improved interfacial adhesion and morphological control are required, i.e., the concept of "compatibilization." Copolymerization into random, block, or graft structures is shown to be a powerful way of dealing with these problems.

REASONS FOR BLENDING

Polymer blending may be viewed as a problem solving technique and, in most cases, the reasons for selecting this approach fall into two categories: property combinations and cost dilution. Most products succeed because of a beneficial combination or balance of properties rather than because of any single characteristic. In addition, a material must have a favorable benefit to cost relation in order to be selected over other materials for a particular application.

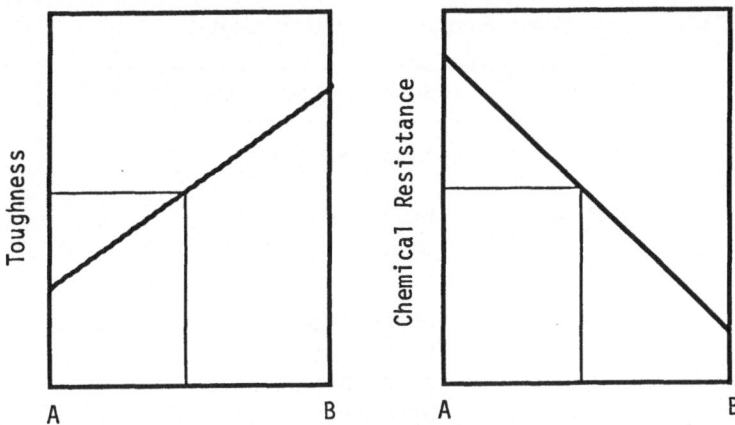

Fig. 1.   Illustration of obtaining beneficial property
           combinations by blending.

The idea of property combinations is best illustrated by an
example. Consider an application that requires a material with
specified levels of chemical resistance and of toughness. A plastic bumper for automobiles is a typical example. The successful
bumper material must be tough enough to survive low speed impacts
and must not fail when contacted by gasoline or oil. Polymer A
may have more than adequate chemical resistance but insufficient
toughness while polymer B is more than tough enough but is lacking
in chemical resistance. If blends of A and B exhibit more or less
additive values of these two characteristics as illustrated in
Figure 1, then certain combinations of these materials may meet
both requirements simultaneously whereas neither one alone can.
Recently, commercial products of this type have been developed
where A is a crystalline, aromatic polyester (e.g., PET or PBT)·
and B is bisphenol-A polycarbonate[28]. In other cases, one might
seek a favorable balance of processibility, flame retardance,
stiffness, thermal resistance, barrier properties, etc.

A particular polymer may have a key deficiency that pre-
cludes expansion of its application into new areas. By blending
this polymer with another that has superior performance in its
area of deficiency, one can sometimes broaden its versatility and,
thus, its market potential. Table 1 lists a number of polymers
that have been the object of intense blend development and some of
the characteristics where improvement is sought.

3

Table 1. Some Frequently Blended Polymers

Polymer	Improvement Sought
PS	toughness (HIPS)
PVC	toughness raise or lower T_g
PPO	processing toughness chemical resistance lower cost
PC	chemical resistance toughness in thick parts
PBT & PET	toughness
ABS	upgrade to engineering thermoplastic level (eg, raise HDT, toughness, chemical resistance)
Nylon	Toughness(dry)

In general, there is a nearly direct relationship between performance and cost among available materials, i.e., high performance materials cost more. This is true because low performance materials that are expensive to produce do not survive. Because of this, some higher performance materials are not able to compete for lower performance markets since there are less expensive materials that will function adequately. Therefore, the producer of a high performance material might seek to broaden market potential through the concept of cost dilution. This is illustrated in Figure 2 where low cost, low performance polymer A and higher cost, higher performance polymer B are blended. If performance is again more or less additive as shown on the left, then a series of lower performance materials may result that can compete with other materials because of the cost dilution factor. Two commercial examples are used to illustrate this principle. The first involves poly(phenylene oxide) as B and polystyrene as

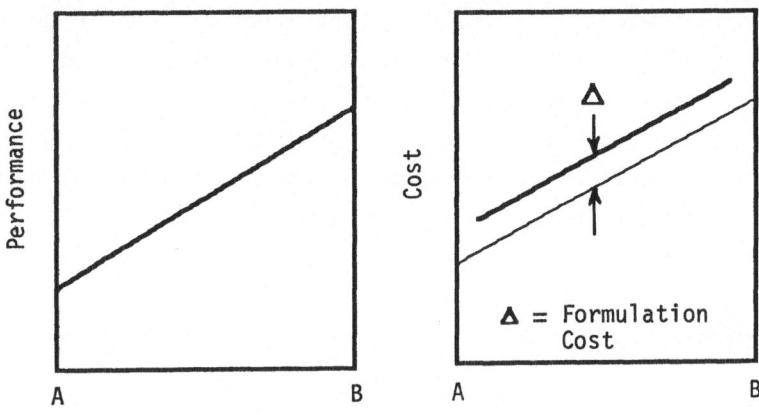

Fig. 2.   Concept of cost dilution

A.   These two amorphous polymers are miscible and their blends
have a single glass transition temperature that progressively increases from that of polystyrene (100°C) up to that of poly-
(phenylene oxide) (~220°).   Thus, blending can give the exact
degree of heat resistance, within these limits, required by the
application without having to pay for more than is needed.   In a
similar way, the cost of polycarbonates can be reduced by blending
with ABS plastics to gain materials with useful performance/cost
ratios.   In both of these examples, some other issues are also
involved.   The processing characteristics of the higher performance poly(phenylene oxide) and polycarbonate are improved by the
addition of polystyrene and ABS, respectively.

    There is an important caveat, however, about the cost of
blends.   Their price cannot be the simple tie line connecting
those for A and B (see right side of Figure 2) since there is a
formulation cost, Δ, that may include extra processing steps,
special additives, and a return on the research investment.

    Based on the above, blending looks like a simple business
where new products can be designed to specification using a computer data base filled with the properties and costs of available
materials.   However, the property relationships used for illustration purposes in Figures 1 and 2 are rarely so simple.   Quite
often, arbitrarily chosen components will, in fact, yield blends
whose key properties are much worse than additive.   That is, the

5

blend components are incompatible. One must select or design components for blending rather carefully to get the desired result. A key issue is the phase behavior or structure of the blend which is governed by how the components "interact" with each other as illustrated in the following sections. Most successful blends including the examples mentioned above are more than simple mixtures casually put together in an extruder. They are the result of sophisticated materials science that may require specially designed components, modifiers and processing protocols.

MISCIBLE BLENDS

When two polymers, A and B, form a homogeneous mixture, many properties of the blend are additive as implied in the discussions of the previous section[13,14]. The problem is that most randomly selected polymer pairs are not thermodynamically miscible; hence, homogeneous mixtures are not obtained. The reasons for this are well-known[13,14]. A necessary thermodynamic condition for miscibility is that the free energy of mixing must be negative and

$$\Delta G_{mix} = \Delta H_{mix} - T \Delta S_{mix} \tag{1}$$

sufficient conditions include satisfying additional stability requirements[13]. When A and B have high molecular weights, the favorable entropy term is very small, if not entirely negligible, while the heat of mixing is generally positive and dominates equation 1. The latter term is often expressed in the following van Laar form where the ϕ_i are volume fractions of the components and

$$\Delta H_{mix} = B \phi_A \phi_B \tag{2}$$

B is the interaction energy density which may also be written in terms of the "chi" notation of the Flory-Huggins theory for polymer mixtures[31-33].

The usual immiscibility of polymeric components can be circumvented if their segments interact in such a way that the heat of mixing is exothermic or B is negative rather than the usual positive value. Negative B values exist when some appropriate specific interaction occurs between the segments of the two polymers, and numerous examples of miscible blends based on this principle have been identified over the past decade.[28] More recently[31-33], it has been shown that intramolecular repulsion between monomer units in random copolymers can be an additional driving force causing them to form miscible blends with other polymers. This notion can be very useful for designing polymeric components for formation of miscible blends and, thus, to achieve multicomponent polymer systems with additive properties.

A simple mean field, binary interaction model[31-33] for copolymer blends gives the following expression for B (equation 2)

$$B = B_{13} \phi'_1 + B_{23} \phi'_2 - B_{12} \phi'_1 \phi'_2 \qquad (3)$$

for mixing of a copolymer of monomers 1 and 2 with homopolymer composed of monomer 3. Here, the ϕ'_i indicate the copolymer composition and the B_{ij} characterize the interactions between monomers i and j in the polymers regardless of their bonding to other units. Even though all binary homopolymer pairs may be immiscible, i.e., all $B_{ij} > 0$, certain copolymer compositions may lead to miscible blends (B < 0) provided the repulsive B_{12} term in equation 3 is strong enough relative to the other contributions. However, if all of the binary interactions are purely dispersive such that the B_{ij} may be calculated from solubility parameter theory[31], exothermic mixing cannot result as demonstrated in Figure 3. The example calculation shows that it is possible by copolymerization to match solubility parameters and have B approach zero for a certain composition. This combined with small departures from solubility parameter theory, or weak nondispersive interactions, can evidently lead to regions of compositions where B < 0 and, thus, windows of miscibility as seen in many homopolymer-copolymer systems.[31-33] An interesting example[34] is styrene/acrylonitrile copolymers containing from 10 to 30% by weight of AN which are miscible with poly(methyl methacrylate), PMMA.

Random copolymerization can be employed in another useful way to control the phase behavior of blends. Many miscible blends exhibit reversible phase separation on heating, i.e., lower critical solution temperature or LCST behavior. In some cases, the phase boundary is below the temperature region necessary for melt processing and on cooling a phase separated mixture persists owing to the slow rate at which homogeneity can be achieved. A good example of this is polycarbonate - PMMA blends[35]. As seen in Figure 4, appropriately prepared blends of these polymers have a single, additive glass transition temperature (left), but the mixture phase separates at temperatures below that needed for melt

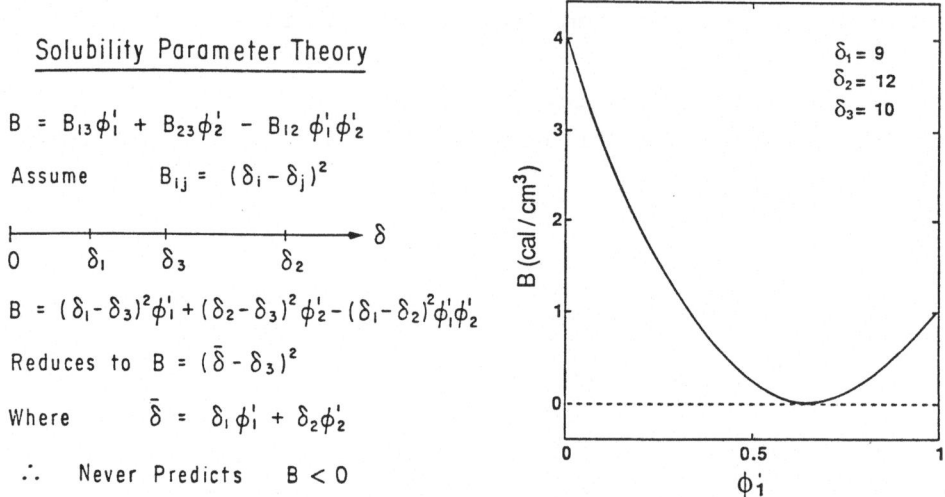

Fig. 3. Calculation of interaction parameters for blends of a homopolymer with a copolymer using solubility parameter theory.

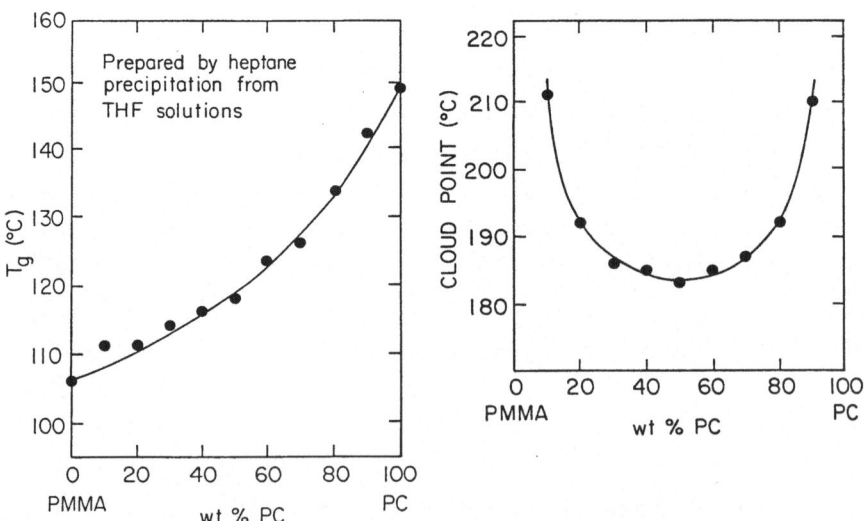

Fig. 4. Phase behavior of blends of poly(methyl methacrylate) and polycarbonate.

processing. Clearly, a method for raising this cloud point curve is needed which amounts to making B in equation 2 more negative. Addition of a comonomer to one of the components has been shown to be a viable approach[36-39] and may, in principle, function by increasing interchain attraction or by increasing intrachain repulsion or both. Some examples are shown below. More complete details for each case are given elsewhere.[36-39]

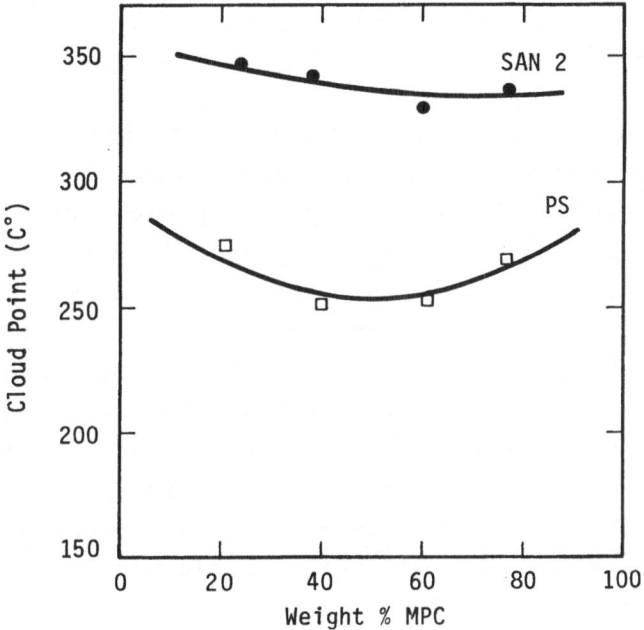

Fig. 5   Cloud point curves for blends of styrenic polymers with tetramethyl bisphenol-A polycarbonate.

Tetramethyl bisphenol-A polycarbonate, MPC, forms miscible blends[36] with polystyrene, PS, that phase separate on heating at the temperature shown in Figure 5. By incorporating only 2% by weight of AN into the styrenic raises this boundary by about 80°C. Higher amounts of AN raise this curve to above the decomposition temperature; however, at 13% or more AN the SAN copolymers are not miscible with MPC. Polystyrene forms miscible blends with poly-(vinyl methyl ether), PVME, that phase separate at quite low temperatures. Copolymerization of very small amounts of acrylic acid with styrene dramatically elevates the phase separation

temperature[38] as seen in Table 2; although, beyond a certain
amount of acrylic acid these copolymers are not miscible with
PVME. Similar but less dramatic results were obtained using
acrylonitrile[38], maleic anhydride[38], or methyl methacrylate[40] as
the comonomer. Returning to the PC-PMMA system, Figure 6 shows
that copolymerizing small amounts of styrene into the acrylic similarly elevates the phase boundary for this system. Equation 3
proves useful for understanding these effects and gives a basis
for designing polymers for controlled miscibility.

IMMISCIBLE BLENDS

Blends of immiscible polymers have complex property-
composition relationships that are rarely additive. Most properties are dramatically influenced by the spatial arrangement of the
phases in the final blend. The morphology is strongly affected by
processing history and can change undesirably during fabrication
steps since it is a dynamic structure. Properties like stiffness,
heat distortion temperature, or barrier behavior are dominated by
the component forming the continous phase and show a sigmoidal
shape versus composition owing to phase inversion. Failure properties and toughness are often much less than additive and may be
inferior to those of either pure component. This feature severely
limits the utility of many potentially valuable blend systems.

The poor mechanical behavior of phase separated blends is
usually the consequence of inadequate adhesion between the phases
that does not allow efficient transfer of stress across this interface. Poor adhesion results from the lack of affinity between

Table 2. Blends of PVME with Styrene/Acrylic Acid Copolymers

Weight % Acrylic Acid	Miscible with PVME?	LCST
0	Yes	~110° C
1	Yes	Above chemical decomposition temperature
3	Yes	Above chemical decomposition temperature
8	No	--

the two polymers that are responsible for their immiscibility. It is clear that the mechanical properties would be more nearly additive if this interfacial zone were strengthened. One way to do this, shown schematically in Figure 7, is to add interfacially active block or graft copolymer "compatibilizers" to the immiscible mixture.[10] In its simplest form, the compatibilizer has block or graft segments which are chemically identical to those in the respective phases, although non-identical segments which are

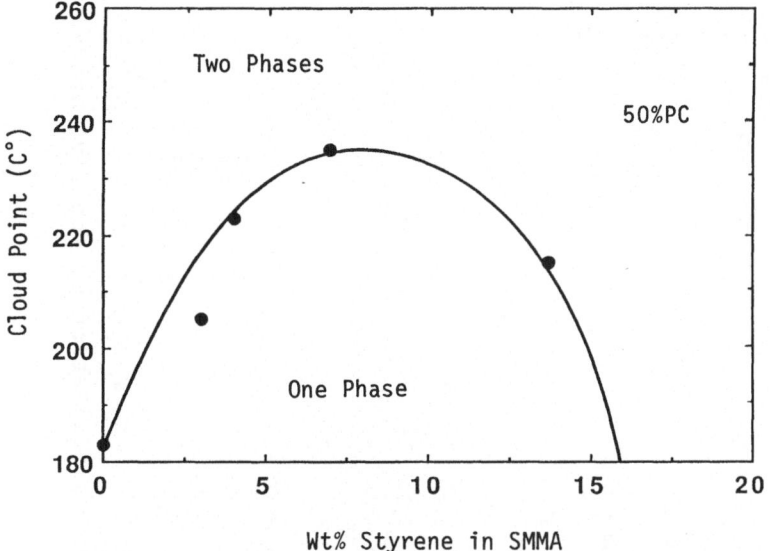

Fig. 6. Effect of styrene comonomer content on temperature of phase separation of polycarbonate blends with methyl methacrylate/styrene copolymers.

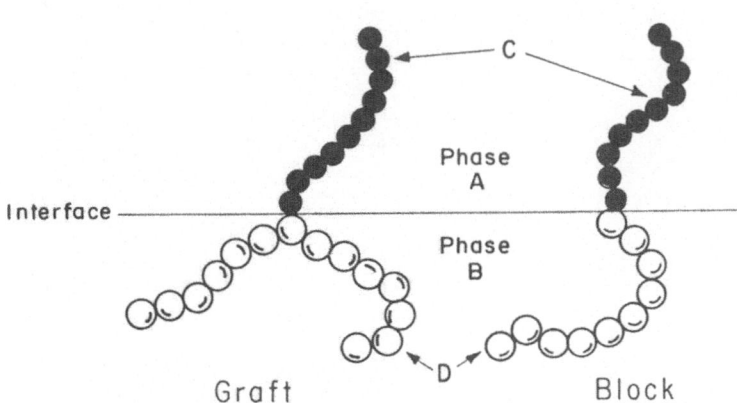

Fig. 7. Compatibilization of immiscible blends with block or graft copolymers at the interface.

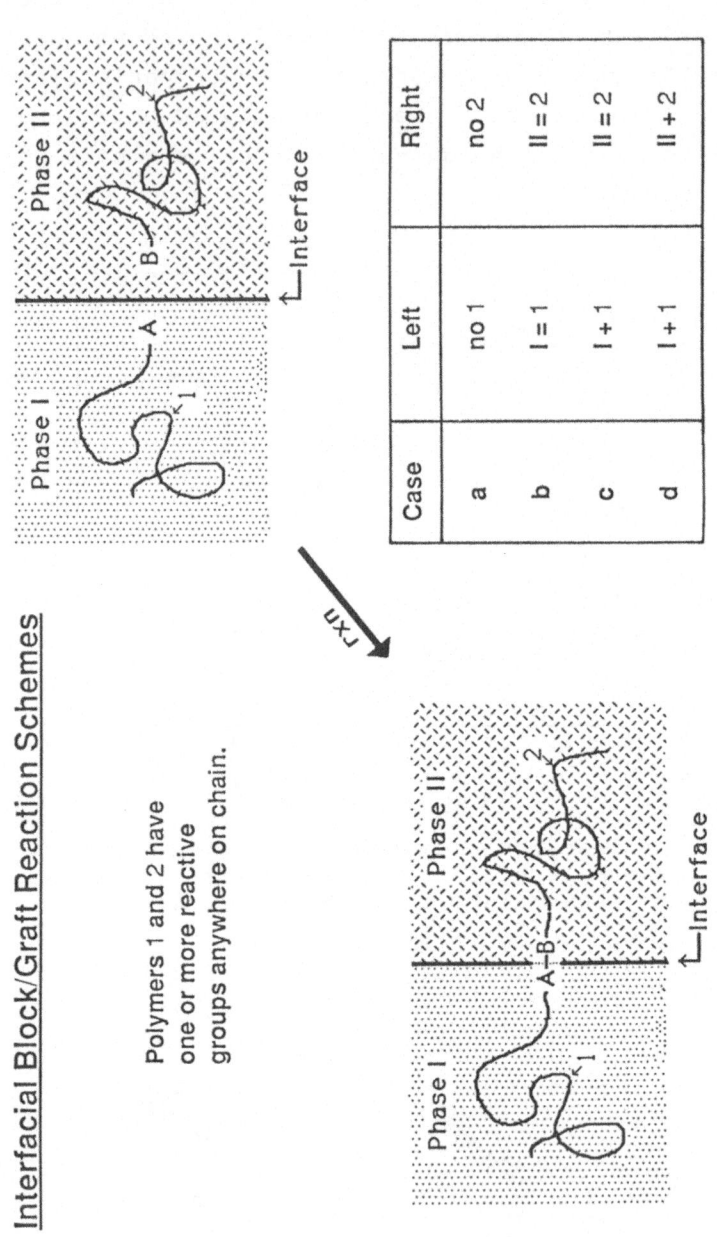

Fig. 8. Compatibilization by reactive processing.

miscible or partially miscible in the respective phases should work equally well to improve interfacial adhesion and blend properties, permit a finer dispersion during mixing, and provide a measure of stability against gross segregation.[41-55]. Some of the most definitive work in this area is that by Teyssie and co-workers[43-49] who have carefully explored compatibilization of immiscible blends containing polyethylene and polystyrene by various hydrogenated butadiene-styrene diblock copolymers. They conclude, based on a combination of microscopy and assessment of mechanical property improvement, that as little as 0.5-2 weight percent addition of diblock is sufficient to achieve a macroscopically homogeneous and stable phase dispersion, that most of the copolymer is at the interface as suggested in Figure 7, that diblock copolymers with blocks of similar size are the most efficient interfacial agents and their activity is effective in a very large range of block copolymer and homopolymer molecular weights for melt blended products, and that diblocks are often more efficient interfacial agents than triblock copolymers and grafts. However, this approach is severely limited by the availability of techniques for preforming appropriate block or graft copolymers. An attractive and generally more useful approach appears to be the formation of such interfacially active species at the interface itself during processing by chemical reactions of functionalized polymer components which are precisely tailored and added in the amount needed. This approach has been described in some recent publications[56-58] and is a part of the rapidly growing field known as "reactive processing."

The idea of in situ formation of block or graft copolymer at the polymer-polymer interface is shown schematically in Figure 8. Phase I and Phase II represent the two immiscible polymers, and the diagram shows a small region from a blend or composite that includes a section of interface; Phase I contains polymer chains of type 1 having one or more functional groups (A) anywhere on the chain while Phase II contains polymer chains of type 2 having one or more different functional groups (B). A and B are chosen so that they will readily react with each other to form a block or graft copolymer as shown. Note that this drawing shows only one functional group per chain which is located at the chain end; however, the concept is not limited to this ideal case. The functionalized polymers 1 and 2 must form homogeneous mixtures (i.e.,

miscible) with Phases I and II, respectively. The nature of the
interfacial zone may have a significant influence on the ease with
which the interfacial reaction can occur. In the limit of a very
sharp interface, the A and B groups can approach each other only
in very restricted ways (see Figure 9) since they will not be able
to penetrate the opposite phase and the probability of reaction
will be limited correspondingly. In the more typical case, the
interface will be somewhat diffuse, and in this zone, segments
from the two phases will intermix creating greater opportunities
for A and B groups to approach each other (as shown schematically
on the right in Figure 9) and subsequently to react.

These approaches to compatibilizing immiscible blends allow
the development of unique products not limited by the immiscibility of the components[59] and offer the opportunity for many
exciting "new" materials.

SUMMARY

In principle, polymer blending is an attractive physical
alternative to new molecules for developing new products. However, to practice this approach successfully requires careful control of phase structure since this affects the properties that can
be achieved. Through appropriate copolymerization methodology one
can manipulate miscibility, phase morphology, and interfacial adhesion to achieve highly engineered blend materials having properties that are additive combinations of the components. In some

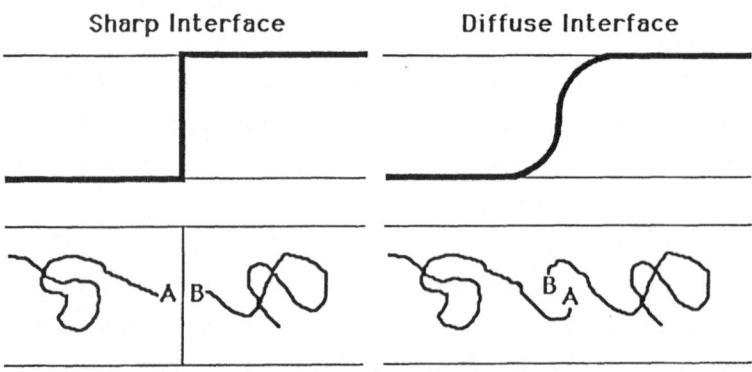

Fig. 9. Reaction configuration at sharp and diffuse interfaces.

cases, individual properties can be better than those for either component, i.e., synergism can occur.

ACKNOWLEDGMENTS

This research was sponsored by the National Science Foundation, Grant No. DMR-86-03131, and by the U.S. Army Research Office.

REFERENCES

1. H. Keskkula, ed., "Polymer Modification of Rubbers and Plastics," Appl. Polym. Symp. 7 (1968).
2. P. F. Bruins, ed., "Polyblends and Composites," Appl. Polym. Symp. 15 (1970).
3. N. A. J. Platzer, ed., "Multicomponent Polymer Systems," Adv. Chem. Ser. 99 (1971).
4. G. E. Molau, ed., "Colloidal and Morphological Behavior of Block and Graft Copolymers," Plenum, New York (1971).
5. L. H. Sperling, ed., "Recent Advances in Polymer Blends, Grafts, and Blocks," Plenum, New York (1974).
6. N. A. J. Platzer, ed., "Copolymers, Polyblends, and Composites," Adv. Chem. Ser. (1975).
7. D. Klempner and K. C. Frisch, eds., "Polymer Alloys: Blends, Blocks, Grafts, and Interpenetrating Networks," Plenum, New York (1977).
8. J. A. Manson and L. H. Sperling, "Polymer Blends and Composites," Plenum, New York (1976).
9. C. B. Bucknall, "Toughened Plastics," Applied Science Publishers, London (1977).
10. D. R. Paul and S. Newman, eds., "Polymer Blends, Vols. I and II," Academic, New York (1978).
11. S. L. Cooper and G. M. Estes, eds., "Multiphase Polymers," Adv. Chem. Ser. 176 (1979)..
12. D. Klempner and K. C. Frisch, eds., "Polymer Alloys II," Plenum, New York (1980).
13. O. Olabisi, L. M. Robeson, and M. T. Shaw, "Polymer-Polymer Miscibility," Academic Press, New York (1979).

14. D. R. Paul and J. W. Barlow, Polymer Blends (or Alloys), _J. Macromol. Sci._ – _Rev. Macromol. Chem._ C18:109, (1980).

15. K. Solc, ed., Polymer Compatibility and Incompatibility: Principles and Practice, Vol. 2 _in_: "MMI Press Symposium Series," Harwood Academic Publishers, Cooper Station, NY (1982)

16. E. Martuscelli, R. Palumbo, and M. Kryszewski, eds., "Polymer Blends: Processing, Morphology, and Processing," Plenum Press, New York (1979).

17. J. W. Barlow and D. R. Paul, _Ann. Rev. Mater. Sci._ 11:229 (1981)

18. I. C. Sanchez, Ann. _Rev. Mater. Sci._ 13:387 (1983).

19. B. J. Schmitt, _Angew. Chem. Int. Ed. Engl._ 18:273 (1979).

20. C. D. Han, "Multiphase Flow in Polymer Processing," Academic Press, New York (1981).

21. O. Olabisi, _in_: "Kirk-Othmer: Encyclopedia of Chemical Technology," 3rd Edition, John Wiley and Sons, Inc., New York, 1982, Vol. 18, p. 443.

22. Preprints for Soc. Plast. Eng. NATEC on "Polymer Alloys, Blends, and Composites," Bal Harbor, FL, Oct. 25-27 (1982).

23. S. Wu, "Polymer Interface and Adhesion," Marcel Dekker, New York (1982).

24. C. D. Han, ed., "Polymer Blends and Composites in Multiphase Systems," _Adv. Chem. Ser._ 206 (1984).

25. Polym. Eng. Sci. 22 (2, 11, and 17) (1982); 23 (11 and 12) (1983); 24 (2, 8, and 17) (1984).

26. D. J. Walsh, J. S. Higgins, and A. Maconnachie, eds., "Polymer Blends and Mixtures," NATO ASI Series, Series E, Applied Sciences, No. 89, Martinus Nijhoff Publishers, Dordrecht (1985).

27. D. R. Paul and L. H. Sperling, eds., "Multicomponent Polymer Materials," _Adv. Chem. Ser._ 211 (1986).

28. D. R. Paul, J. W. Barlow, and H. Keskkula, "Polymer Blends," to appear _in_: Encyclopedia of Polymer Science and Engineering, Wiley-Interscience, New York.

29. T. Alfrey and W. J. Schrenk, "Multicomponent Polymers," _in_: Advanced Technology, P. H. Abelson and M. Dorfman, eds., A Special SCIENCE Compendium, AAAS (1980).

30. A. Dobry and F. Boyer-Kawenoki, _J. Polym. Sci._ 2:90 (1947).

31. D. R. Paul and J. W. Barlow, _Polymer_ 25:487 (1984).

32. R. P. Kambour, J. T. Bendler, and R. C. Bopp, _Macromolecules_ 16:753 (1983).

33. G. Ten Brinke, F. E. Karasz, and W. J. MacKnight, _Macromolecules_ 16:1827 (1983).

34. M. E. Fowler, J. W. Barlow, and D. R. Paul, _Polymer_ 28:1177 (1987).

35. J. S. Chiou, J. W. Barlow, and D. R. Paul, _J. Polym. Sci.: Part B: Polym. Phys._ 25:1459 (1987).

36. A. C. Fernandes, J. W. Barlow, and D. R. Paul, _Polymer_ 27:1788 (1986).

37. K. E. Min and D. R. Paul, _Macromolecules_ 20:2828 (1987).

38. K. E. Min and D. R. Paul, _J. Polym. Sci.: Part B: Polym Phys._, to appear.

39. K. E. Min and D. R. Paul, _J. Appl. Polym. Sci._, to appear

40. Y. Y. Chien, E. M. Pearce, and T. K. Kwei, to be published.

41. J. W. Barlow and D. R. Paul, _Polym. Eng. Sci._ 24:525 (1984).

42. D. R. Paul, _in_: "Thermoplastic Elastomers: Research and Development," N. R. Legge, H. Schroeder, and G. Holden, eds., Carl Hansen Verlag, Munich (1987).

43. R. Fayt, R. Jerome, and Ph. Teyssie, _J. Polym. Sci.: Polym. Lett. Ed._ 19:79 (1981).

44. R. Fayt, R. Jerome, and Ph. Teyssie, _J. Polym. Sci.: Polym. Phys. Ed._ 19:1269 (1981).

45. R. Fayt, R. Jerome, and Ph. Teyssie, _J. Polym. Sci.: Polym. Phys. Ed._ 20:2209 (1982).

46. R. Fayt, P. Hadjiandreou, and Ph. Teyssie, _J. Polym. Sci.: Polym. Chem. Ed._ 23:337 (1985).

47. R. Fayt, R. Jerome, and Ph. Teyssie, _Makromol. Chem._ 187:837 (1986).

48. T. Ouhadi, R. Fayt, R. Jerome, and Ph. Teyssie, _J. Polym. Sci.: Polym. Phys. Ed._ 24:973 (1986).

49. R. Fayt, R. Jerome, and Ph. Teyssie, _J. Polym. Sci.: Polym. Lett. Ed._ 24:25 (1986).

50. D. Heikens, N. Hoen, W. Barentsen, P. Piet, and H. Ladan, _J. Polym. Sci.: Polym. Symp._ 62:309 (1987).

51. W. J. Coumans, D. Heikens, and S. D. Sjoerdsma, _Polymer_ 21:103 (1980).

52. W. J. Coumans and D. Heikens, _Polymer_ 21:957 (1980).

53. S. D. Sjoerdsma, J. Dalmolen, A.C.A.M. Bleijenberg, and D. Heikens, *Polymer* 21:1469 (1980).

54. S. D. Sjoerdsma, A.C.A.M. Bleijenberg, and D. Heikens, *Polymer* 22:619 (1981).

55. R. J. Roe, *J. Chem. Phys.* 62:490 (1975).

56. D. V. Howe and M. D. Wolkowicz, *Polym. Eng. Sci.* 27:1582 (1987).

57. W. E. Baker ;and M. Saleem, *Polymer* 28:2057 (1987).

58. W. E. Baker and M. Saleem, *Polym. Eng. Sci.* 27:1634 (1987).

59. J. M. Heuschen, *in*: "High Performance Polymers: Their Origin and Development," R. B. Seymour and G. S. Kirshenbaum, eds., Elsevier, New York (1986).

FUNCTIONAL USES OF STYRENIC BLOCK COPOLYMER

William P. Gergen

Shell Development Company

Houston, Tx

INTRODUCTION

Styrenic block copolymers are members of the generic family of thermoplastic elastomers.[1-3] They are usually made with a composition which results in a continuous soft-segment phase and thus are usually rubbers. Like most synthetic rubbers they are usually compounded with resins, oils, and fillers to produce commercial compounds that can be processed on ordinary thermoplastic processing equipment above the thermal transitions of the components of the compound. The artifacts produced usually have the characteristics of a rubber artifact; tires, hoses, shoe-soles, gaskets, rubber bands, electrical cords, to name a few. These represent uses as monolithic structural thermoplastic elastomers. Most TPE's share this identity with conventionally vulcanized rubbers. It is the intended and primary use of this class of materials.[4]

The functional uses of styrenic block copolymers are usually a result of the unique characteristics of these types of synthetic polymers, related to the nature of the block segments covalently linked in the polymer chain; to the nature of the energetic interactions between the chain segments, or to the network stucture that is achieved in combinations where the block copolymer has multiple hard blocks in each chain. The functional utility of the block copolymer is to provide a specialized and vital function to the whole without imparting the "rubber" identity to the whole. For example, the rubber seals in a power transmission device are vital to the operation of the transmission, yet there is no identity imparted to the transmission by the rubber gaskets and seals. A rubber which is added to a thermoplastic or thermoset to provide a mechanism of impact resistance has a functional use rather than a monolithic structural use even though that is the function of the plastic.

Some of the interesting and commercially significant uses of styrenic block copolymers have as their intended function:

1. To provide a mechanism of gellation for petroleum-based greases and encapsulants.[5] Related to this, block copolymers provide viscosity index improvement in multi-grade lubricating oils.[6] Tri-block copolymers

form continuous gel networks and di-block copolymers form micelles in petroleum base-stocks depending on the interaction between the block copolymer segments and the oil (solvent). Above a critical temperature, or below a critical concentration, the ordered structure disappears and a homogeneous solution is formed where viscosity is dependent on the dissolved block copolymer concentration and the temperature.[7]

2. Tri-block copolymers are used to improve cold flow, increase softening point, and increase low temperature impact strength to asphalts.[8]

3. Block copolymers provide impact strength in conventional high-impact polystyrene interpolymers.[9-11] The phase structure can be precisely controlled by selective solvation to form micelles. An optimum size and volume of the phase inclusions can be accomplished. In melt blending, styrene-diene block polymers can form disperse impact absorbing phases in polystyrene and polyolefins polymers.

4. They provide an amphiphillic function in blends of dissimilar polymers, where favorable interactions exist between one of the homopolymers and one of the block copolymer segments.[12] Hydrogenated block copolymers which reduce surface tension in polystyrene-polyolefin blends is one example. Another example is the use of a derivatized block copolymer, such as terminally carboxylated styrene-diene di-block copolymers, to provide a surfactant function for the dispersion of inorganic fillers in thermoplastics or thermosets.

(We will discuss in some detail the following two functional uses.)

5. The use of styrenic block polymers to provide a critical level of cohesive strength or integrity for melt forming operations. The specific uses involve the "green" processing of tire rubbers and the thermoforming of crystalline thermoplastics.

6. The use of styrenic block polymers to provide a topological template in producing co-continuous (IPN) blends of thermoplastics.[13]

As we describe these topics we shall include in our discussion a brief description of the synthesis, morphology, and properties of styrenic block copolymers, especially as they relate to their uniqueness for the particular functional use.

SYNTHESIS AND MORPHOLOGY

Styrenic block copolymers are made by anionic living polymerization using sec-butyl lithium as a preferred initiator in non polar solvents such as cyclohexane or toluene.[14-16] In the normal sequential polymerization, the initiator reacts with a molecule of styrene to form a polystyryol lithium species which then propagates with transfer of the initiator anion to the active chain end.

$$RLi \ + \ xS \ \text{---} \ RSxLi$$

Ideally there are no termination or crossover ractions; the degree of polymerization is dependent on the monomer to initiator ratio and the molecular weight distribution is very close to unity. The polystyrol lithium can then initiate the reaction of a second monomer, isoprene or butadiene most commonly, using isoprene as an example:

$$RSxLi + yI \ \text{---} \ RSxIyLi$$

After the second monomer is polymerized, the active chain anion is used to initate the third block monomer, commonly styrene again:

$$RSxIyLi + zS \quad --- \quad RSxIySzLi$$

The process could be continued to form a linear multi-block copolymer or terminated by adding methanol to the solution. Alternatively, instead of initiating a third block as was shown above, the di-block species, RSxIyLi, could be reacted with a di-functional coupling agent to form a tri-block copolymer with a doubling of the isoprene block molecular weight. This copolymer would have identical terminal blocks which is very difficult to achieve sequentially.

$$RSxIyLi + R'X_2 \quad --- \quad (RSxIy)_2R' + 2LiX$$

Star block copolymers can also be formed by reaction of the di-block species with multi-functional coupling agents such as polydivinyl-benzene and polychlorosilanes. In any coupling reaction we introduce the possibility of leaving un-coupled di-block in the polymer which is incapable of contributing to the strength of the polymer. With sequential polymers, the terminal polystyrene block can be substantially different in molecular weight from the first block resulting in a poorer quality network structure. These polymers, as di-blocks, tri-blocks, or star block polymers, make up the first generation unsaturated styrenic block copolymers. The newer generation, the saturated block copolymers, are made by solution hydrogenation of the first generation type polymers or of variants of these polymers.

Hydrogenation of 1,4 polymerized polyisoprene leads to a perfectly alternating ethylene-propylene rubber. Such a polymer is non-crystalline, has a Tg of about -60°C, and makes a good rubber. On the other hand, hydrogenation of 1,4 polymerized polybutadiene leads to a polyethylene (polymethylene) chain which is crystalline and not a good rubber at all. Butadiene also readily polymerizes through the 1,2 carbons which are converted to a butylene structure on hydrogenation. The crystallinity introduced by the methylene enchainment of a 1,4 polymerized butadiene monomer can be broken up by the butylene structure. About 40% 1,2 addition is needed to ensure a low level of crystallinity and preserve a low Tg.[17]

THE ORDERED, DISORDERED AND MELT STATES

Block copolymers obtain a high cohesive strength from a virtual or physical network resulting from microphase separation. There is a repulsive interaction between the chains of each of the segments, characterized by the Flory-Huggins interaction parameter, which forces a spatial density fluctuation emanating from each A-B junction. There is a long range order with periodic structure that is detectable in electron-micrographs and by various scattering techniques. Domains of pure A or B segments are formed, separated by an interphase whose size is dependent on the degree of mixing of segments at any temperature.[18] At low concentrations of the A segment, domains of A are spherical and arranged on a body-centered cubic latice. As the concentration increases the spherical domains become cylinders which are hexagonally packed. At even higher concentrations the domains become two-dimensionally continuous alternating lamellae of A and B.[19] As concentration in B decreases the domain structure inverts and B domains show the same series of geometries.

In Figure 1, the density of the A and B segements of an A-B block copolymer with alternating lamellar morphology is plotted as it is scanned along the direction of the periodic structure, r. The dotted line shows the density distribution of the B monomer, isoprene in this case and the solid line the density distribution of the A monomer, styrene. The interface region thickness is on the order of 20 A and the domains are on the order of hundreds of angstroms in the direction normal to the domain surface.[20]

The entropy associated with placement of the junctions is reduced by the stricture of the A-B junctions at the interface, while the conformational entropy is reduced by the restriction of chain motions within the confines of the domain boundaries.[21] This is the ordered state and is characteristic of di-blocks as well as network-forming block copolymers. This ordered state can be detected by observing the periodicity of the A-domain-interphase-B-domain stucture by the scattering of x-rays, light, or neutrons. The order-disorder transition temperature, Tc, is measured by these techniques.[22]

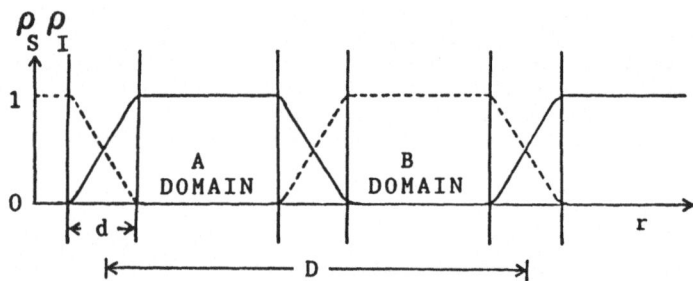

Figure 1. Segmental density fluctuation of alternating lamellae of pure polystyrene-b-polyisoprene-b-polystyrene block copolymer. The density (referenced to each pure segment composition) varies between 0 and 1 for each domain phase and in the interphase region which has finite thickness, d. The periodic pattern distance of the lamellae structure is denoted by D.

The participation of a few end-block chains in more than one domain can form a network in the ordered state which is functionally equivalent to a chemically crosslinked network with much lower dissociation temperature or failure activation energy.[23] The domains are physically large compared to molecular dimensions, on the order of several hundred angstroms, and provide also a reinforcing or stiffening effect.

In most cases, the temperature region of the ordered state also contains the "melt" state where thermoplastic processing can occur. Because of the liquid-crystal-like order, the viscosity of the block copolymer is usually high and is non-Newtonian with reference to dependance on shear rate. As the repulsive interaction energy or the Flory-Huggins interaction parameter increases, the temperature dependence of viscosity decreases. For styrene-diene polymers, the activation energy of flow in the melt state is similar to that of polystyrene.[24] As the interaction gets smaller the distinction between melt state and disordered state disappears.

At some elevated temperature, the entropy gain is sufficient to overcome the repulsive interaction of the chain segments and the polymer becomes molecularly mixed. This is the disordered state in the block polymer. This occurs (providing the segments have a reasonable molecular weight) at a temperature far above the glass transition temperature of the hard phase, ie. polystyrene. If the repulsive interaction is sufficiently energetic, as is the case for some hydrogenated block polymers, the order-disorder temperature, can be very high, on the order of 350°C.[25]

VULCANIZABLE BLOCK COPOLYMERS

Natural Rubber has always been a strategic material but especially so in time of war. After the second world war began, the US government built and operated several SBR plants to provide an alternative to Natural Rubber for tires. After the war these plants were purchased by members of the chemical industry and the U.S. synthetic rubber industry was born. The thrust of the research done in this industry was the pursuit of better synthetic tire rubbers to replace natural rubber. SBR did not seem to be the answer. Synthetic polyisoprene was made first by alkyl lithium polymerization and introduced commercially in the late 1950's. At the same time it was discovered by new NMR techniques that natural rubber consisted of 100% cis 1,4 polyisoprene, while the alkyl lithium polymerized synthetic polyisoprene averaged about 94% cis 1,4 at best with the remainder being 3,4 polyisoprene. Natural rubber showed a great advantage over synthetic polyisoprene in rubber processing and tire building operations; it was able to crystallize on stretching even in the un-cured state and provided sufficiently high strength to be handled without flowing under load. Dr. N.R. Legge, 1987 Charles Goodyear medalist and, at that time, director of the synthetic polyisoprene program at Shell Development Company, recalled that it was in attempting to provide a solution to this problem of poor "green strength" in alkyl lithium polymerized polyisoprene that the styrenic block copolymers were first synthesized.[26] The functional use of the first block copolymers was not as an identifiable monolithic rubber structure but provided a vital function to another identifiable material and lost its identity in this process.

Block copolymers as they are made today are not suitable for tire rubbers even when vulcanized because of the excessive hysteretic losses present at temperatures approaching the polystyrene glass transition. These losses contribute to heat buildup in the tire and consequent loss in properties such as strength and tear resistance. Since the losses increase with temperature below 100°C the rate of heat generation also increases with temperature, a very unstable situation for tires. This situation also dooms the use of polystyrene-b-polyisoprene-b-polystyrene polymers with low end-block concentrations. In terms of other physical properties, there is no compelling reason to believe that vulcanized block copolymers would be inferior to other tire rubbers. What limits their consideration here is most likely the higher cost of the block copolymer synthesis.

One way to approach the problem with the hysteretic loss associated with the onset of the polystyrene glass transition in vulcanized S-I-S polymers is with block copolymers having small random styrene-isoprene copolymer end-blocks, and a pure polyisoprene center block, IS-I-IS. The copolymer composition is chosen to produce a Tg of 40-45°C, at about 60% styrene content. At temperatures experienced in "green" processing or tire building the block copolymer would be below Tg and strong, while at the elevated temperatures typically experienced by the vulcanized product in actual use, the end-blocks would be above their glass transition temperature and could not display the spiraling hystersis of the styrene block copolymers. This situation could be even better if the vulcanization of the rubber were accomplished over a sufficiently long period above the order-disorder transition temperature, Tc, of the IS-I-IS block copolymer. In such a case the rubber would have a permanent structure that contained no phase-separated block segments at all but would incorporate the end-blocks into the bulk of the polymer by difusion.

Hashimoto, in comparing a tapered styrene-isoprene copolymer (47% styrene, 50K Mol weight) to the same composition and molecular weight of pure styrene-b-isoprene polymer with equal block sizes, found a value of Tc of 150°C for the tapered polymer and 250°C for the pure polymer.[27] One could expect the Tc to be even lower in the case where the end-blocks were uniform random copolymer and the center block was isoprene of much higher molecular weight. The equation for the temperature dependence of the interaction parameter is:

$$\chi = A_1 + B_1/T$$

Hashimoto found in the disordered state for a pure block copolymer of isoprene and styrene by analysis of scattering intensity fluctuation:

$$\chi = -0.0937 + 66/T$$

The order-disorder transition will be lowered in a block copolymer with both A and B monomers in the end block according to:

$$(A_1 + B_1/T_r) \ (f_{AS} - f_{BS})^2 = A_1 + B_1/T_{ro}$$

where: f_{AS}, f_{BS} are concentrations of styrene in the A and B blocks;

and T_{ro} is the order-disorder transition temperature, Tr, for the pure block, in this case a $(S-I)_2$ polymer.

Evaluation of the above for a block copolymer with uniform random SI copolymer end-blocks with 60% styrene, leads to an estimate of Tr of about 90°C, well below the typical vulcanization temperatures of 145-160°C. The indication is that the proposed scheme of developing a temporary network which dissociates during the final vulcanization is feasable. This situation would be the epitome of a truely functional use of block polymers since, in this case, the block copolymer structure itself as well as its identity cease to exist after the function has been accomplished.

The target polymer to accomplish this scheme is a block copolymer with random isoprene-co-styrene end-blocks, constituting about 10% of the total polymer weight, with a pure cis 1,4 polyisoprene center block. Normally, high molecular weights (>500K) of the 1,4 polyisoprene are required for adequate rubber processing but in this case adequate viscosity for milling and processing can be supplied by the block copolymer network structure. A reasonable target molecular weight is then set by a convenient end-block molecular weight, about 20K.

The reaction scheme requires that the end block be polymerized first in cyclohexane and then transfered as living polymer to a isopentane solution to make the second block of pure cis 1,4 polyisoprene. To complete the systhesis the di-block polymer is coupled with a di-ester or di-functional chlorosilane coupling agent to obtain the final linear tri-block copolymer. The series of reactions are as follows:

Random copolymer end-block:

$$RLi + xS + yI \ \text{---} \ R(SI)x+yLi$$

Polyisoprene center block:

$$R(SI)x+yLi + zI \ \text{---} \ R(SI)x+yIzLi$$

Coupled linear tri-block:

$$R(SI)x+yIzLi + R'X_2 \ \text{---} \ [R(SI)x+yIz]_2R' + 2LiX$$

When the monomers are added simultaneously with the initiator in the first step solvent, the copolymer formed is rather severely tapered from pure polyisoprene to pure polystyrene, because of the difference in reactivity of the isoprene and styrene monomers. Such a tapered copolymer is equivalent to a block copolymer with block segments rich in isoprene and block segments rich in styrene. The structure of the entire polymer after second step polymerization looking just at segment junctions in or near the interphase region is consequently:

$$(polysoprene-b-polystyrene-b-polyisoprene)_2$$

This polymer is probably incapable of forming a domain network structure at room temperature. Another problem with the tapered copolymer is that it will not remain in solution in the second step solvent and the second step reaction will begin as a heterogeneous reaction with somewhat unpredictable consequences. A uniformly random copolymer can be synthesized by programmed addition of the more reactive monomer, isoprene, during the first step polymerization. Although the reactivity of styrene can be considerably increased by the addition of polar co-solvents such as THF, the polar solvent would seriously degrade the micro-structure of the second-step polyisoprene segment.

Figure 2 shows the spatial density fluctuation for a block copolymer with a uniform random styrene-isoprene endblock and one with a severely tapered end block. In the tapered copolymer end-block, a second periodic structure is created within the long period of the uniform end-block structure because of the repulsive interaction of segments within the end-block polymer itself. The domain structure of such a block copolymer is unknown, it would be unlikely to form a strong network. This point is confirmed in the data shown in Figure 3 which shows the stress strain curves of the block copolymers along with other examples of green rubbers in a typical HAF-black loaded tread compound. The successful green processing of rubbers relates to the stress strain response and specifically to the work done in deforming the green compound. "Green Strength" is usually expressed as the work function obtained by integrating the area under the stress-strain curve expressed in Joules/cm^3. The values of the work function for the tapered end-block copolymer is 0.5 compared to a value of 22.5 J/cc for the uniformly random end-block copolymer. The shape of the curve for this polymer is preferable to the curve displayed by the SBR tread compound since it displays the characteristic of self-reinforcment similar to that of natural rubber.

The vulcanized properties of a number of tread rubber candidates including the block copolymers, compounded in a HAF tread compound are shown in Table 1.

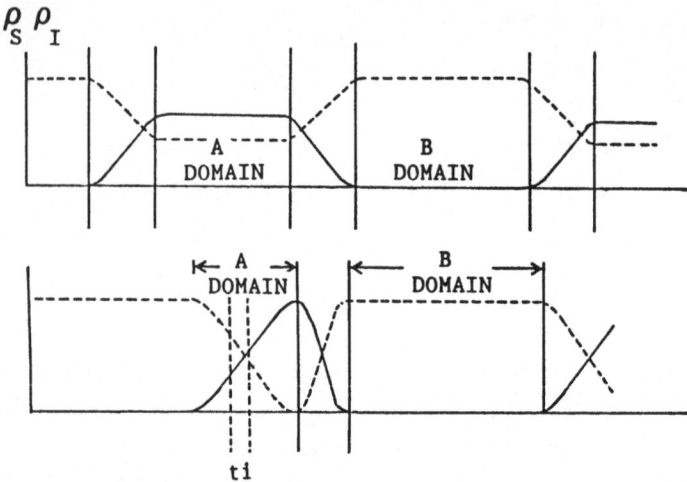

Figure 2. Segmental density distribution of block copolymers with styrene-isoprene copolymer random and tapered terminal blocks segments. The upper figure is for a random styrene–isoprene copolymer end-block with uniform composition. The lower figure is that of a severely tapered copolymer end block. The A domain is viewed as equivalent to a block copolymer which is itself microphase-separated with its own interphase region.

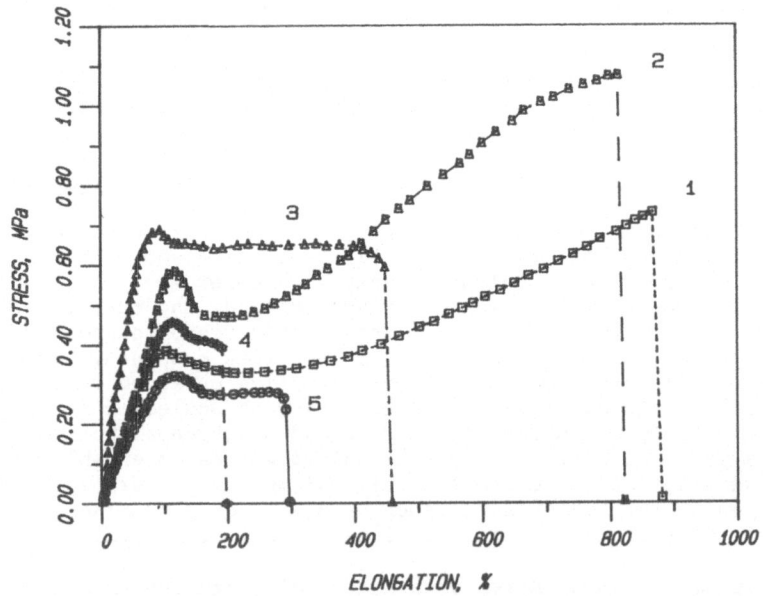

Figure 3. Stress–strain response of "Green" tire tread compounds. Samples compounded with HAF black and tested at 5.0 cm/min. using ASTM D specimens. Samples are: (1) Natural Rubber #1RSS, (2) IS–I–IS block copolymer with random copolymer end-blocks 60% styrene, (3) SBR (23% styrene) (4) lithium polyisoprene rubber (5) high-cis Ziegler polyisoprene rubber.

Alkyl lithium polyisoprene ceased to be manufactured in the U.S. in the early 1970's in favor of polyisoprene polymerized by Ziegler catalysis and because of the greatly improved domestic reserve of natural rubber.[26] Nevertheless, the work that was carried out on the copolymer end-block tire rubbers is a worthwhile addition to our knowledge. Although the idea and the work with this polymer occurred over a decade ago, new insight into the nature of ordered and disordered states and their transition makes this subject relevant in todays world.

Table 1.   Physical Properties of Vulcanized Rubber Tread Compounds

Rubber Type	NMR Cis 1,4	Flexometer Temp Rise T°C @70~C	Tensile Strength 23°C MPa	Tensile Strength 120°C MPa	Tear Strength 120°C Pli
1. N.R. #1RSS	100	17	24.8	16.9	350
2. (IS-I)2	90	21	23.5	15.2	350
3. SBR 1500		32	18.8	10.2	190
4. Alkyllithium PI	89	22	24.1	12.1	250
5. Ziegler PI	98	19	26.2	16.9	380
6. (S-I)2	88	47	21.8	11.9	200

Compounded with 50 phr HAF, 5 phr aromatic oil, sulfur and accelerators.
Cured 30 minutes 150°C in a closed mold.
Samples numbered in accordance with Figure 3.

THERMOFORMABLE CRYSTALLINE RESINS

A use of block copolymers which is is very relevant in current technology of thermoplastic processing deals with a problem that is common to all crystalline thermoplastics and especially to molecular weight limited condensation polymers such as nylons and polyester. A normal thermoplastic process such as injection molding is based on developing a fluid polymer mass which can be transferred into a mold with a maximum speed and minimumm pressure, both of which can be accomplished with low molecular weight, and then allowing the material to set up to the desired shape by some solidification process, in this case by crystallization caused by supercooling. The kinetics of crystallization are improved also by low molecular weight. When the process requires some deformation in the melt state such as is the case in thermoforming operation, the polymer which is being processed must have a reasonable consistency and strength above the melting point, just exactly opposite of the previous case. This is the normal characteristic of amorphous polymers which depend on kinetic transition phenomona, such as Tg or rapidly increasing viscosity with cessation of shear (non-Newtonian flow), to achieve solidification sufficient for demolding or melt deformation.[28,29] With polyolefins and particularly with polypropylene, the problem can be assisted considerably by increasing the molecular weight and broadening somewhat the molecular weight distribution but an acceptable balance between rate of crystallization and melt strength in order to achieve a reasonable processing window is difficult to achieve. The problem is similar in nature to the green-strength problem in tire rubbers. Polypropylene lacked sufficient strength in processing to achieve acceptable performance in thermoforming so a device to increase strength in melt processing was needed. Subsequently, the solution had no relevance to the function of the solidified object, unless it coincidently provided a useful property modification.

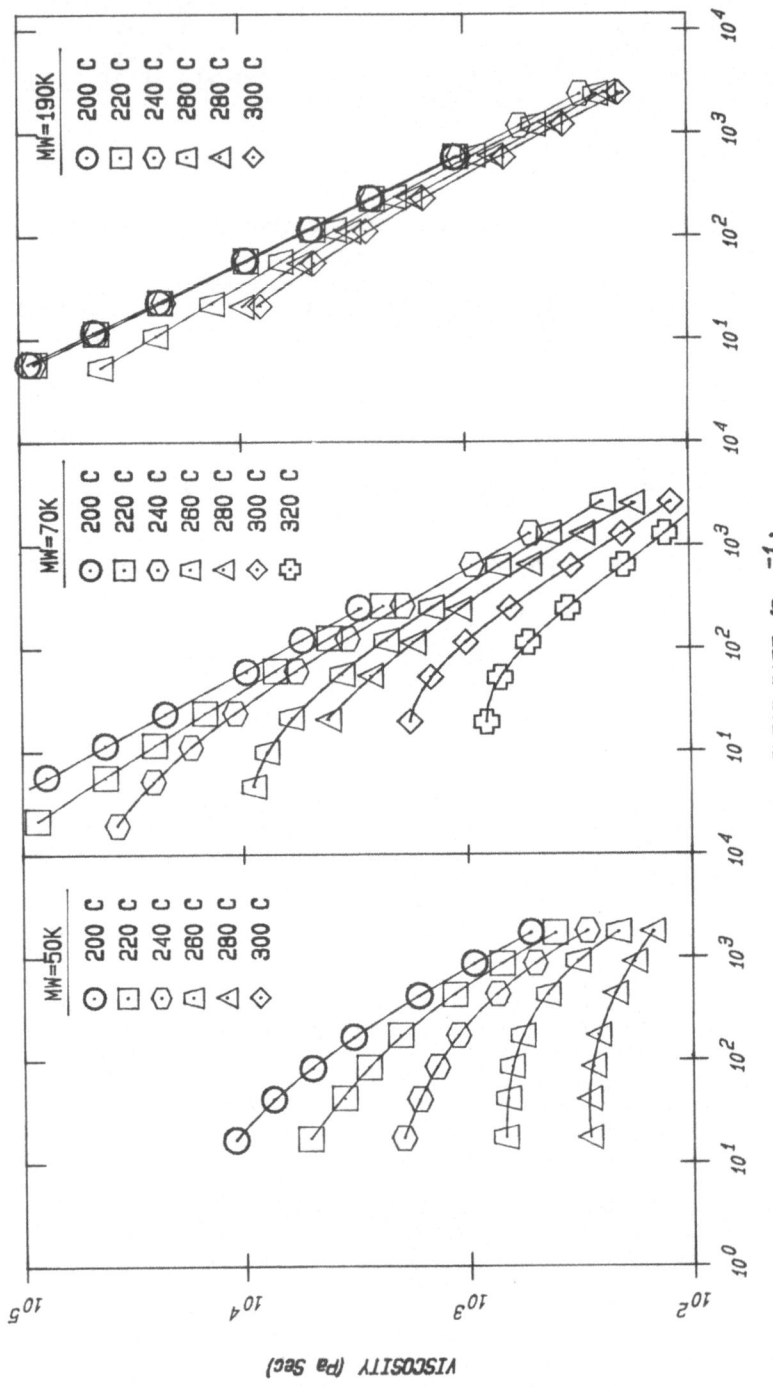

Figure 4. Viscosity as a function of shear-rate for three S-EB-S polymers of various molecular weight in the "melt" state. Melt-network structure is apparent in the set for the highest molecular weight polymers which have a slope of nearly –1 and where viscosity is relatively insensitive to temperature. The curves for the polymer on the left show the polymer is likely going through a disorder transition.

Again, block copolymers proved to be the answer, but this time as an additive to the polypropylene thermoplastic, and additionly, this time involving saturated block copolymers made by hydrogenation of styrene-diene copolymers. When the hydrogenated block copolymers were first synthesized, unlike the unsaturated block copolymers they were observed to be nearly intractable at molecular weights that were at all useful from the point of view of developing properties. It was discovered that this problem of poor flow could be resolved in most cases by adding moderate amounts of polypropylene resin. These blends could then be processed very similarly to the polypropylene homopolymer. It was found that the addition of very small amounts of the propylene polymer reduced the measurable melt viscosity by as much as two orders of magnitude. A theory of "shear-plane lubrication" was formulated to explain the phenomonum. An identical phenomonum has been observed in other difficult-to-process polymers. We will discuss this point further in the next section.

When intensive mixing was used to make these blends, for instance internal mixers such as Banbury mixers or twin-screw extruders, it was discovered that molded shapes from the blend material had an unusual morphology. The block copolymer rubber phase could be completely, even analytically extracted from the artifact leaving a residue which retained the shape and volume of the original artifact. What was left was an open-celled spongue with very small cell dimensions dependent on the volume fraction of the original rubber phase. This suggested that the block copolymer and polypropylene resin combination existed as co-continuous phases in the volume of the blend artifact. Over a period of time, other evidence such as stained STEM (scanning transmission electron microscopy) confirmed this morphology which we have come to accept now as a consistent characteristic of these blends. We found also that other polyolefins; polyethylene, polybutylene, and polymethylpentene also showed the same phenomonum but to a lesser extent than polypropylene.

Since we knew that the blends produced a co-continuous morphology which was detectable in the solid, it was likely that this structure was also present in the melt provided the particular block copolymer had a significantly higher order-disorder transition than the melt processing of polypropylene required and sufficiently energetic mixing was used to make the blends. Consideration of the viscosity-shear rate curves measured by a capillary rheometer, as shown in Figure 4., for several hydrogenated block copolymers, (polystyrene-b-ethylene-co-butylene-b-polystyrene), showed that at polypropylene blend mixing temperatures, about 250°C, the lowest molecular weight S-EB-S polymer was above or approaching the order-disorder transition since the viscosity, which was comparatively low, was tending toward Newtonian behavior at low shear rates. On the other hand the highest molecular weight S-EB-S showed very little temperature dependence at all. The viscosity of the high molecular weight polymer, at the lower temperatures at least, appears to obey:

$$\eta = \gamma^{-1}$$

indicating that not only is the periodic long range order intact, but there is in addition a melt network structure in the block copolymer itself which exhibits a finite yield stress.[30,31] This is an extremely important point for the stability of the co-contiuous phases in blends of dissimilar materials as we will discuss a little later.

Blends containing 5.0% and 7.0% of this high molecular weight S-EB-S polymer were thremoformable within the same window as thermoformable ABS between 170°C and 200°C. A comparison of the complex dynamic viscosity at 0.1 radian/sec for: (1) a thermoformable ABS grade, (2) a polypropylene copolymer and (3) two blends containing the high molecular weight S-EB-S is shown in Figure 5. The dotted lines indicate the minimum value of complex viscosity where the thermoformaing process can be accomplished. The ABS has an adequate processing window (in temperature), the polypropylene copolymer an extremely narrow one, and the two blends of PP and S-EB-S have windows equalivalent to that of the ABS polymer. The melt network structure of the S-EB-S assures adequate viscosity above Tm. In addition, the blend containing the block copolymer, which was based on polypropylene homopolymer, had a notched Izod Impact strength of 0.43 KJoules/Meter of notch, compared with a value of 0.25 KJ/M for the polypropylene copolymer. The function of the block copolymer in this case was two-fold, to provide the necessary melt characteristics for thermoformability and to provide a toughening mechanism for the rather brittle polypropylene homopolymer.[32]

THERMOPLASTIC IPNS

The nature of the co-continuous phases in the polypropylene S-EB-S blends leads to a generalization of the specific function of the block copolymer in systems like this. The now-classical model of this characteristic phase structure is that of the chemical IPNs where each network is actually a phase which is constrained or condensed by a mechanism which has infinite relaxation time and where three dimensional spatial continuity of each phase is achieved. This is topologically identical to those IPNs which are achieved by sequential polymerization and crosslinking within a pre-existing network structure or the

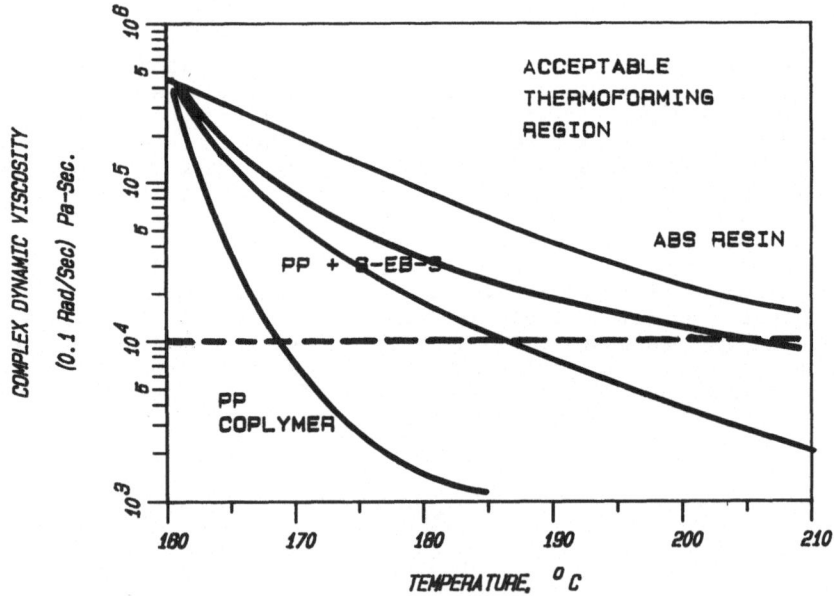

Figure 5. Viscosity as a function of temperature in thermoforming resins. Dotted line shows minimum viscosity for automated thermoforming operations. ABS is acceptable, polypropylene is not. The melt network derived from the saturated block copolymer S-EB-S imparts good thermoformability to polypropylene.

simultaneous polymerization of two disimilar monomers by non-interacting crosslinking mechanisms. The essential features of the thermoplastic IPNs are the three-dimensional spatial continuity (contiguous phases), and an energetic barrier to phase retraction or demixing. Also required is a method of synthesizing the IPNs from dissimilar polymers.

The advantages of IPNs derive from their three dimensional morphology. Other types of polymer blends of two polymer species include homogeneous or miscible blends where there is a strong interaction between the constituents of the polymers which results in a negative free energy of mixing, and matrix-disperse blends where the higher viscosity component is dispersed in the lower viscosity matrix with various levels of adhesion between particle and matrix. The IPNs are the third kind of morphology. The expression of the properties of the minor constituent component is possible in the case of miscible blends where a new property set may be realized and in the case of the IPNs where the contribution is in proportion to the fractional volume and the level of that property which is manifest in that polymer species. The properties resulting from the IPN structure can be crudely approximated by linear additivity mixing rules while this is not the case for matrix disperse blends. As an example, an IPN blend of equal volumes of polypropylene and polycarbonate which have flexural modulus of 1.55 GPa and 2.34 GPa respectively at room temperature, and containing the highest molecular weight hydrogenated block copolymer (190K), at a level of 10%V and which has a modulus of 0.06 GPa, obeys a logarithmic mixing rule with exponent of 0.5 with respect to flexural modulus:[33]

$$M_{Bl}{}^{0.5} = M_a{}^{0.5}Vf_a + M_b{}^{0.5}Vf_b + M_c{}^{0.5}Vf_c$$

Where: a,b,c are polypropylene, polyester, ans S-EB-S polymer constituents

$M_{a,b,c}$ are the moduli of the constituents

$Vf_{a,b,c}$ are the volume fractions of the constituents

The blend has a modulus of 1.63 GPa, confirmed by experiment. The estimate from simple mixing rule would be 1.75 GPa. On the other hand a matrix disperse blend in which the polcarbonate and the S-EB-S polymer were dispersed as uniform particles in a polypropylene matrix would be expected to follow a parallel model and if we presume that the S-EB-S does not participate in the modulus, the value calculated is close to the above 1.6 GPa. There doesn't seem to be an advantage for the IPN structure. However, if the temperature were to be increased to 120°C, the polypropylene component would have a modulus of only 0.14 GPa, the polycarbonate a modulus of 2.2 GPa, and the S-EB-S a modulus of 0.06 GPa. The blend would have a modulus of 0.93 GPa, about 135 Ksi, whereas the matrix disperse blend would have a modulus of about that of the polypropylene, 0.15 GPa or 22 Ksi. At this temperature there is nearly a decade higher modulus that is attributable to the IPN structure over the matrix-disperse structure. The modulus of the three components and the IPN blend is shown as a function of temperature in Figure 6. The level of modulus corresponding to the HDT at 264 psi stress is shown by the dotted line. The IPN structure is helpful in significantly raising the HDT temperature over that of polypropylene in these polypropylene-polycarbonate blends.

All crystalline thermoplastics; nylon, polyester, polyacetal, as well as polypropylene, display a similarly large loss in modulus as temperature is increased. As a consequence they are not able to perform under stress at even slightly elevated temperatures. The answer to this problem is found in IPN blends of the crystalline resin with amorphous resins which do not suffer the same loss in modulus with temperature. This is one of the strongest reasons that the current commercial slate of engineering resin blends are alloys, usually with co-continuous morphology, of crystalline and amorphous polymers.

The function of the block copolymer, especially in IPN blends which have more than two components, (ternary or higher IPNs), is to provide the directing or templating function which can be interpenetrated by the more mobile phases, and to provide a barrier to phase retraction.

In many cases where the nature of the constituents is not grossly different, the IPN morphology can be achieved by isoviscous mixing within a narrow compositional range. If the response of two polymers to a given shear field is sufficiently similar, intermixing of the phases can occur without losing connectivity especially if both components have the necessary melt strength to sustain the deformations imposed during the mixing process. In these cases the characteristic cell size of the IPN which is one measure of the state of interdispersion of the components, is uniformly dependent on the Flory-Huggins interaction parameter between the components in the blend, or between the blend constituents and the EB block of the S-EB-S polymer. This is reasonable since this energetic repulsion is the source of the interfacial tension resisting the mixing of the components. The phase size can be reduced by increasing the intensity of the shear field but only to a limited degree before degradation or other limiting phenomona occur. This situation is observed even with the addition of low molecular weight S-EB-S which must be used if viscosity matching is adhered to.

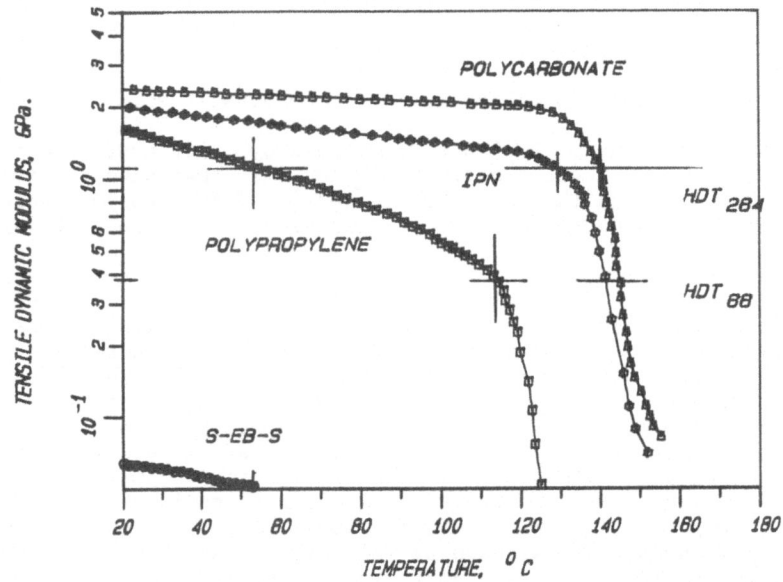

Figure 6: Modulus-temperature relationships for a ternary IPN blend and its constituents, polycarbonate, polypropylene, and S-EB-S block copolymer. The modulus corresponding to deformation in ASTM HDT test calculated from linear elasticity is shown at applied stress of 264 and 66 psi. are shown bye crossed lines.

32

On the other hand, we find that in binary mixtures, when the high molecular weight S-EB-S polymer is one of the components of the IPN, the resulting phase size is a maximum of 1 micron regardless of the nature of the other polymer or polymers. Phase size is often less than 1 micron in cases where the blending component is a polyolefin and has low concentration in the blend. We find also that the IPN structure is the preferred structure over all volume concentrations above approximately 7 %v of the S-EB-S polymer regardless again, of the nature or the viscosity of the blended polymers. We are fairly certain that the mechanism underlying this tendency toward IPN formation and "shear plane lubrication" are manifestations of the same phenomonum. Since this S-EB-S polymer is so highly energetically ordered in a network structure in the melt, true viscous flow does not occur. Instead, in the presence of the interpenetrating resin phase, (which by definition must be much lower in viscosity), the network fragments along shear planes into "flow units" which are consistent in size with the demands of the shear field. Because of its high strength, the network can be deformed by the interpenetrations of the other resin without losing connectivity or without permanently exceeding a minimum size under the balance of forces present. In addition, the fragmented network is capable of reconnecting by the diffusion of a very few chain segments during the process. The high molecular S-EB-S polymer dominates the topology and energetics of the dynamic mixing process and in the solid state always has a characteristic morphology which we call "skeletal"; bulky, curved, and everywhere convex, while the remaining phase or phases are concave and appear to be the mirrored complement, or "antitropic", to the S-EB-S phase.

If the thermoplastic IPN is formed without the functional use of a block copolymer, it is usually transient at some specific melt temperature. For instance, IPNs with phase size of 0.1 micron can be made by mixing 70% polybutylene and 30% polypropylene at 230°C and quenching. When this IPN is held in a closed mold for periods up to 1 week, the characteristic cell size increases gradually to over 20 microns. Connectivity is retained but the clear indication is that eventually gross retraction would occur. When the same blend contained 6% of the high molecular weight S-EB-S is made and tested there is no measurable phase size growth. This we attribute to the finite yield stress of the continuous S-EB-S melt network which easily resists the energetic repulsion due to the chemical potential difference of the blend constituents. In fact, the same function can be accomplished by blend components which by themselves exhibit a shear yield stress.

In blends of three or more polymers, the morphologies that can exist within the IPN definition are numerous depending on the pairing of constituents. For example, in a blend of A, thermoplastic polyester, B, polypropylene, and C, S-EB-S polymer, we have found that the morphologies which can exist and have been observed have the following characteristics:

- the three components can exist as co-continuous phases with an alignment of ACB so that there are no A-B contacts, (the C phase would be an appropriatly sized interphase between A and B).

- the components can each be continuous and spatialy free to contact each of the other two phases,

- an intimate IPN can form with characteristic phase size between C and B which forms a second IPN structure with the A phase at another characteristic cell size

- A can become disperse in B and together form an IPN with C.

We calculate that the number of possible variants of the morphology of ternary blends with two co-continuous phases is twelve and with three co-continuous phases is nine.

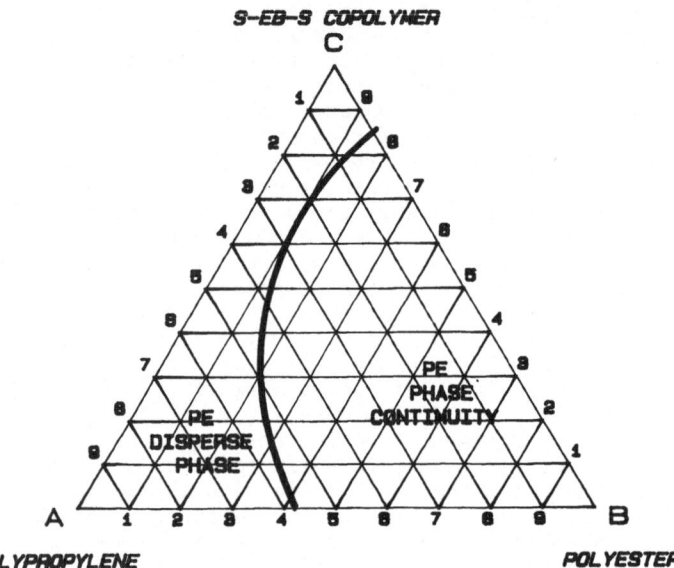

Figure 7. The "Continuity Cliff" in a IPN blend of PP, PBT, and S-EB-S. All compositions to the right of the line have continuous PBT phase and co-continuous PP and S-EB-S phases. Compositions to the left of the line have disperse PBT phases and co-continuous PP and S-EB-S phases.

The complexity of structure inherent in the higher order IPN blends sometimes leads to unexpected results which do not achieve the targeted morphology and properties. The above blend system is one example of such a situation. Very fine cell size IPN blends were made with thermoplastic polyester and the high molecular weight S-EB-S polymer. The logical extension of this work was to add polypropylene to achieve a practical and economical blend. The logic behind this blend required a continuous polyester phase to achieve stuctural integrity above the Tm for the polypropylene phase. What we found is that when the targeted composition, richer in polypropylene than the other components was made, the polyester phase collapsed, ie. lost continuity. Examining this system in some detail revealed the phase diagram shown in Figure 7 which clearly shows a "continuity cliff" located in the most desirable composition range. At 50% polypropylene, 30% S-EB-S was required to achieve continuity in the system with the remaining 20% polyester. This is an econmically poor solution to the problem of high temperature structural integrity. The polyester cell size was found to depend on the concentration of the S-EB-S polymer in the blend going from a size of >30 micron at 10% S-EB-S to 1 micron at 60% S-EB-S. The S-EB-S in this system greatly favored the PP phase to the distinct disadvantage of the IPN forming capability with the more-polar polyester phase.

We believe that the ultimate solution to this problem and related problems with other IPN blend combinations lies in the next generation of styrenic block copolymers which will provide chemical groups on the hydrocarbon backbone which will provide specific interactions with complementary groups in a number of polar polymers and will succeed in significantly lowering interfacial tension between the component of the IPN blends. These polymers have been synthesized and are now in the development stage.

References

1. G.Holden in "Block and Graft Polymerization",(R.I.Ceresa, Ed.), Vol 1,Chapter 6,Wiley,New York,1973.
2. S.L.Aggawal,Ed.,"Block Polymers",Plenum Press(1970)
3. L.H.Sperling,Ed.,"Recent Advances in Polymer Blends, Grafts and Blocks",Plenum,New York,1973
4. N.R.Legge,G.Holden,H.E.Schroeder,Eds.,"Thermoplastic Elastomers", Chapter 13,Carl Hanser,Munich,1987
5. R.J.Roe,M.Fishkis and C.J.Chang,Macromolecules,14,1091(1981)
6. K.M.Hong and J.Noolandi,Macromolecules 14,727(1981)
7. G.Kraus in "Block Copolymers: Science and Technology",D.J.Meier,Ed., MMI press.,Gordon Breach Sci. Publisher(1983)
8. J.H.Collins and W.J.Mikols, paper presented at the 60th Meeting of the Association of Asphalt Paving Technologists,San Antonio,Tx,Feb.1985
9. G.E.Molau,Ed.,"Colloidal and Morphological Behavior of Block and Graft Copolymers",Plenum Press,(1971)
10. A.L.Bull and G.Holden,J. Elastomers and Plastics 9,281(1977)
11. G.Riess in "Thermoplastic Elastomers",(N.R.Legge,G.Holden,H.E.Schroeder, Eds.),Chapter 12
12. D.R.Paul and S.Newman,Eds.,"Polymer Blends",Vol. 2,Chapter 12, Academic Press,New York,(1978)
13. W.P.Gergen,R.G.Lutz,S.Davison in "Thermoplastic Elastomers",(N.R.Legge, G.Holden,H.E.Schroeder,Eds.),Chapter 14
14. M.Morton,"Anionic Polymerization: Principles and Practice",Academic Press,New York,(1983)
15. J.E.McGrath,Ed.,"Anionic Polymerization, Kinetics Mechanics and Synthesis",ACS Symposium Series No. 166,ACS,Washington,D.C.(1981)
16. L.J.Fetters and M.Morton,Macromolecules 2,190(1969)
17. W.P.Gergen,paper presented at the ACS Rubber Division Meeting, Los Angeles,Ca.,May 1985
18. D.J.Meier in "Thermoplastic Elastomers",(N.R.Legge,G.Holden, H.E.Schroeder,Eds.),Chapter 11
19. L.Leibler,Macromolecules,13,1602(1980)
20. T.Hashimoto,M.Shibayama and H.Kawai,Macromolecules,16,1093(1983)
21. S.Krause,Macromolecules 11(1971)
22. T.Hashimoto in "Thermoplastic Elastomers",(N.R.Legge,G.Holden, H.E.Schroeder,Eds.),Chapter12/3
23. P.-G.deGennes,J. Physics,(Paris),36,1199(1975)
24. G.Holden,E.T.Bishop and N.R.Legge,"Thermoplastic Elastomers",Proc. International Rubber Conference,1967,MacLaren and Sons,London(1968)
25. T.Hashimoto,K.Nagatoshi,A.Todo,H.Hasegawa and H.Kawai, Macromolecules,7,364(1974)
26. N.R.Legge in "Thermoplastic Elastomers",(N.R.Legge,G.Holden, H.E.Schroeder,Eds.),Introduction

27. K.Mori,H.Hasegawa and T.Hashimoto,Polym. J.,17,799(1985)
28. J.N.Hay in "Flow-Induced Crystallization",(R.L.Miller,Ed.), MMI Monograph Vol 6,p.69,Gordon and Breach Sci,London(1979)
29. S.Middleman,"Fundamentals of Polymer Processing",Chapter 14, McGraw-Hill,New York(1977)
30. K.R.Arnold and D.J.Meier,J. Appl. Polym. Sci.,14,427(1970)
31. W.P.Gergen,R.G.Lutz,S.Davison in "Thermoplastic Elastomers", (N.R.Legge,G.Holden, H.E.Schroeder, Eds.), Chapter 14
32. R.L.Danforth and P.S.Byrd, paper presented at the Detroit meeting of the SPE,(RETECH),November,1982

MISCIBILITY ENHANCEMENT VIA IONIC INTERACTIONS IN THE PHENYLATED POLYPHENYLENE OXIDE/POLY(ETHYL ACRYLATE) BLEND SYSTEM

R. Murali[+] and A. Eisenberg*
Department of Chemistry
McGill University
801 Sherbrooke Street West,
Montreal, P.Q. H3A 2K6, Canada

F. W. Harris
Institute of Polymer Science
University of Akron
Akron, Ohio, U.S.A.

R. K. Gupta
Daychem Laboratories
Dayton, Ohio, U.S.A.

ABSTRACT

Blends of sulfonated phenylated polyphenylene oxides with ethyl acrylate-4-vinyl pyridine copolymers were prepared by solution blending. Dynamic mechanical measurements were used primarily to study miscibility. A blend of unfunctionalized phenylated polyphenylene oxide with pure poly(ethyl acrylate) is immiscible, as seen from a two-step descent in the storage modulus curve and the presence of two loss tangent peaks. A dramatic improvement in miscibility is seen from the presence of a single but broad descent in the storage modulus curve for stoichiometric blends of functionalized polymer mixtures of equivalent weight $\leq$ 600. While the breadth of the drop in the storage modulus curve with temperature suggests that the mixture is heterogeneous, the absence of two well defined steps in the storage modulus curve indicates that these blends exhibit the effects of strong interactions. Proton transfer from the sulfonic acid to the 4-vinyl pyridine ring has been confirmed by infrared study.

INTRODUCTION

The area of polymer blends has been a very active one in

Present address: + BFGoodrich Chemical Company
                   Avon Lake, Ohio 44012, U.S.A.

the recent past, because of enormous improvements in properties which can be achieved in some cases as a result of blending.[1-8] Most polymer pairs are immiscible, or at best only partly miscible. Thus, methods of enhancing miscibility have also been under investigation. Among the techniques found most useful are specific interactions such as dipole-dipole[9], hydrogen bonding[10-12], charge (electron) transfer[13,14], ion-dipole[15], ion-ion[16-18] and metal complexation[19,20]. The two methods involving ionic interactions have been the subject of special interest in this laboratory. Several examples which are of relevance to the present work are reviewed briefly below.

In order to enhance the miscibility of the otherwise immiscible polystyrene (PS)/poly(ethyl acrylate) (PEA) blend, Smith and Eisenberg[16] incorporated small amounts of interacting groups such as styrene sulfonic acid (SSH) in the PS chains, and a similar quantity of a proton acceptor, 4-vinyl pyridine (4-VP), in the PEA chains. For the sample containing 2% of interacting groups, two loss tangent maxima at -3° and 89°C were found. Even though the compatibility appeared to have increased (as judged from a slight decrease of the slope of the G' vs. T in the glass transition regions in dynamic mechanical studies), the absence of a single glass transition temperature and the opacity of the sample indicated that the polymers were not completely miscible. Complete miscibility was found only when the interacting group content was 4% or higher. For example, a single glass transition temperature of 72°C was found for a blend sample containing 5% of interacting groups. This value lies between the Tg's of the salts of the precursor polymers, i.e., 124° and 4°C for the sodium salt of the PS copolymer and the PEA copolymer quaternized with methyl iodide, respectively. Furthermore, the samples were found to be transparent. Proton transfer from the SSH to the 4-VP rings, resulting in the formation of ion pairs, was evident from FTIR studies. Thus, miscibility was achieved through ionic interactions between the sulfonate (SO_3^-) and vinyl pyridinium (VPH^+) groups. A similar approach was used by Zhou and Eisenberg[17] to enhance the miscibility of PS/poly-isoprene blend. Complete miscibility was again found for samples containing 5% or more of interacting groups.

Natanshon and Eisenberg[21] investigated the interactions between PS and poly(methyl methacrylate) (PMMA) containing ion pairs of the type discussed above by ^{1}H NMR spectroscopy in solution at 85°C. They found that proton transfer from the SSH group to the 4-VP ring was very rapid, but that the alignment of two polymer chains was a slow process. The latter was monitored by the upfield shift of most of the proton signals of the PMMA copolymer, but primarily that of the methoxy protons, which was due to the screening effect of the neighboring aromatic styrene rings. For a blend containing 11% of interacting groups, the chain alignment or coil overlap process reached a steady state value only after ca. 2 h. This study showed that mixing, even in solution, is slow, and suggested that the mixing conditions are perhaps even more important in ionomer blends than has been found for other systems.

Poly(phenylene oxides) have been used as one component in blends because of their excellent physio-chemical properties. The best known example is that based on poly(styrene) and poly-(2,6-dimethy-1, 4-phenylene oxide) (PPO), sold under the trade name Noryl by General Electric[22]. The addition of PPO to polystyrene raises the glass transition, improves mechanical properties and improves resistance to flammability. Detailed studies on the dynamic mechanical and dielectric properties of these blends and several similar systems have been performed by MacKnight et al[23,24], among others.

Pugh and Percec[14] successfully accomplished the miscibility of otherwise immiscible PPO and poly (epichlorohydrin) pairs by introducing either donor (carbazolyl) or acceptor (3,5-dinitrobenzoyl) pendant units connected to the polymer chains through methylene spacer groups.

In the present study, miscibility behavior is explored in the blend system containing a high Tg component, phenylated polyphenylene oxide, and a low Tg component, poly(ethyl acrylate). The structure of phenylated polyphenylene oxide is given in Figure 1; for the sake of brevity the polymer is designated as Ø P Ø O. The unfunctionalized blends are explored first, followed by an investigation in which the sulfonated Ø P Ø O's (S Ø P Ø O - Acid) are mixed with ethyl acrylate copolymers containing a proton acceptor group, 4-vinyl pyridine. The primary purpose of this study is to see whether the ionic interactions are sufficient to induce also miscibility

in this highly dissimilar system in which the glass transition
temperatures of the components differ by ca. 300°C.

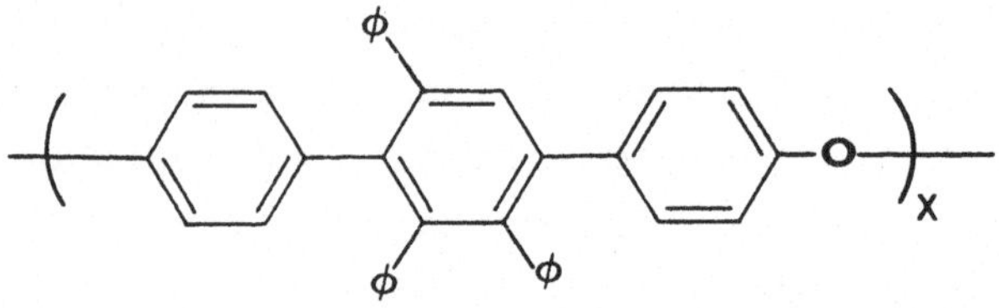

Figure 1.   Structure of phenylated polyphenylene oxide
(∅ P ∅ O)

EXPERIMENTAL

Poly (ethyl acrylate) (PEA), as well as ethyl acrylate copolymers containing varying concentrations of 4-vinyl pyridine
(1960 < equivalent weight (EW) < 630), were prepared in
connection with other projects[16,17,25], and were generously
provided by D. Duchesne. The polymerization procedure and
physical properties of these materials have been reported
elsewhere[25].

One additional ethyl acrylate copolymer (PEA-4VP) sample
containing 21 mol% 4-vinyl pyridine (480 EW) was synthesized
for the present study. The synthesis was performed in solution
by free radical copolymerization[25]. Ethyl acrylate (EA) and
4-vinyl pyridine (4-VP) monomers were purified by fractional
distillation under reduced pressure and were subsequently
stored under N_2 atmosphere at -10°C prior to use. 216g of
EA, 19g of 4-VP, 1.2g of benzoyl peroxide (BZ_2O_2) and 350g
of benzene were used in the reaction. The polymerization was
performed at 56°C for 6-1/2 hours and the conversion was kept
low so that heterogeneity in the material was less than
30%[25]. The copolymer was recovered by precipitation into a
tenfold (volume) excess of mixed hexanes.

All copolymer samples were dried initially under vacuum at room temperature for one week to remove most of the residual hexanes, and were later kept in a vacuum oven at 60°C until constant weight was reached. The 4-vinyl pyridine contents were determined potentiometrically using perchloric acid[26]. The viscosity average molecular weights of polymers prepared under identical conditions were found to be between 200,000 and 300,000[25].

The phenylated polyphenylene oxide (Ø P Ø O) was sulfonated using a slightly modified procedure of Makowski et al[27]. A detailed description of the sulfonation and some properties of these materials is reported elsewhere[28]. Sulfonated polymers (S Ø P Ø O Acid) with equivalent weights of 1200, 600 and 510 were used for blending with ethyl acrylate copolymers.

Two blending procedures were employed. In the first, the S Ø P Ø O-Acid and the ethyl acrylate copolymers were separately dissolved in 1,4-dioxane (2% w/v) at room temperature. After complete dissolution, the ethyl acrylate copolymer solutions were added slowly to the vigorously stirred sulfonated polymer solutions at room temperature. In the case of the stoichiometric S Ø P Ø O-Acid (600 EW)/PEA-4VP (630 EW) and S Ø P Ø O Acid (510 EW)/PEA-4VP (480 EW) blends, a gel-like precipitate was obtained during the addition of the ethyl acrylate copolymer solutions. The mixture was further stirred for 3 hours prior to the freeze-drying step. The freeze-dried samples were kept in a vacuum oven initially at 60°C for one week, then further dried at 180°C for a few days prior to use.

In the second procedure, after separate dissolution of the macromolecules in dioxane, the ethyl acrylate copolymer solutions were added slowly to the vigorously stirred sulfonated polymer solutions at 90°C instead of room temperature as in the first procedure. The mixture was further stirred for 3 hours at 90°C. Blend samples were recovered using the same procedure as in the first method.

Blending of the Ø P Ø O and PEA homopolymers was performed at room temperature in the same manner as that used for the ionic blends. However, a clear solution was obtained during blending of the homopolymers.

A Nicolet 7000 FTIR instrument was used for the infrared studies with 120 scans accumulated for all samples. The materials used for the FTIR study were prepared as follows: samples of polymers of low glass transition temperatures which were soluble in THF were prepared by the deposition of a polymer solution onto an NaCl window. For moderately high glass transition temperature polymers soluble in THF, samples were prepared by solution casting onto a glass plate. Thin films of polymer samples insoluble in THF were compression molded at 150° to 200°C (depending on the sample). An FTIR calibration curve was obtained from ethyl acrylate copolymers which had previously been neutralized completely with per-chloric acid. This curve was used to determine semiquanti-tively the percent proton transfer in the blends.

Dynamic mechanical studies were performed using a computer-ized torsional pendulum from -100° to 300°C at frequencies varying from 4Hz for the glassy region to 0.1 Hz for the low modulus region. A heating rate of ca. 0.8°C/min. was employed during the experiment. The blend samples used for the torsion-al pendulum studies were prepared by compression molding between 150° and 225°C, using a maximum pressure of 45 MPa for 15 minutes.

Blend samples are designated by their polymer equivalent weights.

RESULTS

The glass transition temperatures of the ethyl acrylate copolymers and of the sulfonated Ø P Ø O are reported else-where[25,28], but are summarized in Table 1 for convenience.

The dynamic storage modulus (G') and loss tangent (tan δ) as a function of temperature of the pure PEA and Ø P Ø O homopoly-mers and of the Ø P Ø O / PEA (50/50 w/w) blend are shown in Figures 2 and 3. A typical two-step descent of the storage modulus (where each step corresponds to the T_g of the indivi-dual polymer component) is observed for the unfunctionalized Ø P Ø O/PEA system (Figure 2). This behavior is typical of an immiscible blend. Immiscibility is also evident from the presence of two peaks in the loss tangent curve (Figure 3).

Table 1.   Glass Transition Temperatures of Copolymers

Material	mol%	EW	Tg by DSC+ (°C)	Tg T.P.* (°C)
PEA-4VP	0		-11	-16
	5.1	1960	- 7	- 8
	10.5	952	- 6	0
	16.0	630	11	10
	21.0	480	16	20
S Ø P Ø O - Acid	0		287	280
	39.3	1200++	328	331
	78.7	600	356	N.D.
	92.6	510	357	N.D.

+ heating rate of 20°C/min
* heating rate of ca. 1°C/min
++ EW's based on a repeat unit containing 6 phenyl rings and one oxygen atom   (See Figure 1)
N.D. values were not determined due to decomposition of the samples during molding.

It should be noted that the Tg of the PEA component in the blend sample is shifted to a slightly higher temperature and the Tg of the Ø P Ø O component is moved to a lower temperature by comparison with glass transition temperatures of the pure homopolymers.

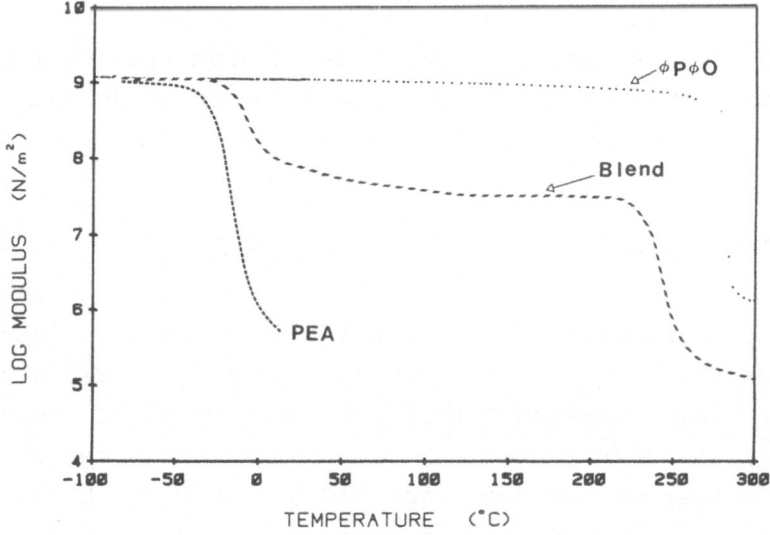

Figure 2.   Variation of the storage modulus (G') with temperature for PEA, Ø P Ø O and a blend of the two (50/ 50 w/w).   Experimental points are only shown for Ø P Ø O.

Similar results were obtained for a 50:50 (w:w) blend of Ø P Ø 0 homopolymer with PEA-4VP (630 EW) (Figures not shown), where again a typical two-step descent of the storage modulus and two loss tangent peaks are seen. All these samples are opaque. Naturally, the PEA-4VP (630 EW) copolymer has a higher Tg value than the pure PEA. This increase in the Tg value is also seen in the blend.

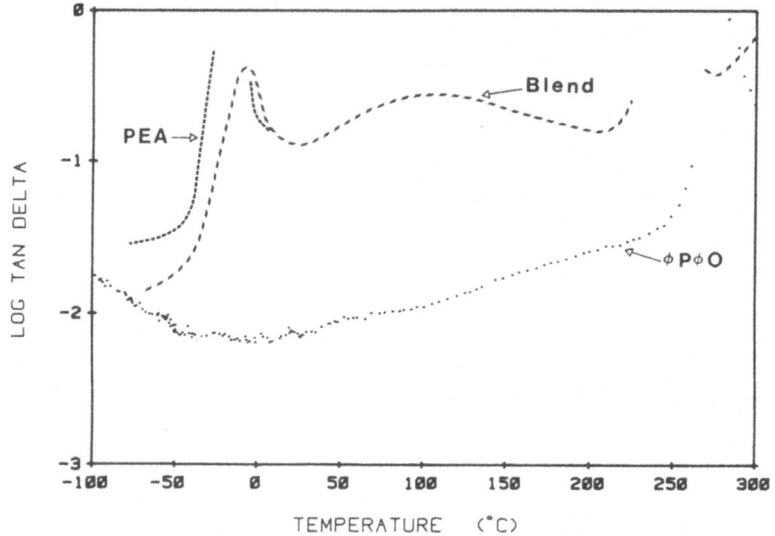

Figure 3. Variation of the loss tangent (tan δ) with temperature for PEA, Ø P Ø 0 and a blend of the two (50/50 w/w). Experimental points are only shown for Ø P Ø 0.

Use of sulfonated Ø P Ø 0 results in a dramatic improvement in miscibility between the S Ø P Ø 0 - Acid (600 EW) and PEA-4VP (630 EW). This is evident from the shape of the G' curve (Figure 4) for that system, in that the two-step descent for the immiscible systems has been replaced by a single, albeit broad descent for the stoichiometric blend of S Ø P Ø 0 - Acid (600 EW)/PEA-4VP (630 EW).

Miscibility enhancement is also evident from the dramatic
decrease in the intensity of the tan δ peak of the ethyl acrylate component (Figure 5). As expected, a non-stoichiometric
blend of S Ø P Ø O - Acid (600 EW)/PEA-4VP (630 EW) (27/73 w/w)

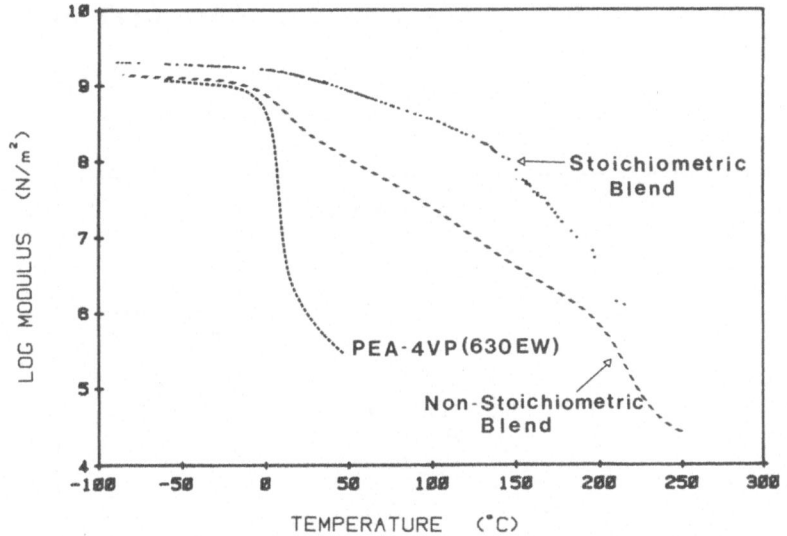

Figure 4. Variation of the storage modulus (G') with
temperature for PEA-4VP (630 EW), a stoichiometric blend
of S Ø P Ø O - Acid (600 EW)/ PEA-4VP (630 EW) (48.8/51.2
w/w) and a nonstoichiometric blend of S Ø P Ø O - Acid
(600 EW) / PEA-4VP (630 EW) (27/73 w/w). Experimental
points are only shown for a stoichiometric blend of
S Ø P Ø O - Acid (600 EW)/PEA-4VP (630 EW) (49/51 w/w).

exhibits a lower degree of miscibility enhancement, which is
clearly apparent from the reappearance of the two-step descent
of the storage modulus as well as the presence of a strong tan δ
peak due to the ethyl acrylate component (Figures 4 and 5).

Figures 6 and 7 show the effect of variations in the contents of the interacting groups on miscibility of stoichiometric blends. The S Ø P Ø 0 - Acid (1200 EW)/PEA-4VP (1960 EW)
(38/62 w/w) and S Ø P Ø 0 - Acid (1200 EW)/PEA-4VP (952 EW)
(55/45 w/w) (not shown) blends exhibit a two-step decrease in
the storage modulus curve. By contrast, the S Ø P Ø 0 - Acid
(600 EW)/PEA-4VP (630 EW) (49/51 w/w) and S Ø P Ø 0 - Acid (510
EW)/PEA-4VP (480 EW) (52/48 w/w) blends exhibit only a single
but broad descent in the storage modulus, which will be dis-

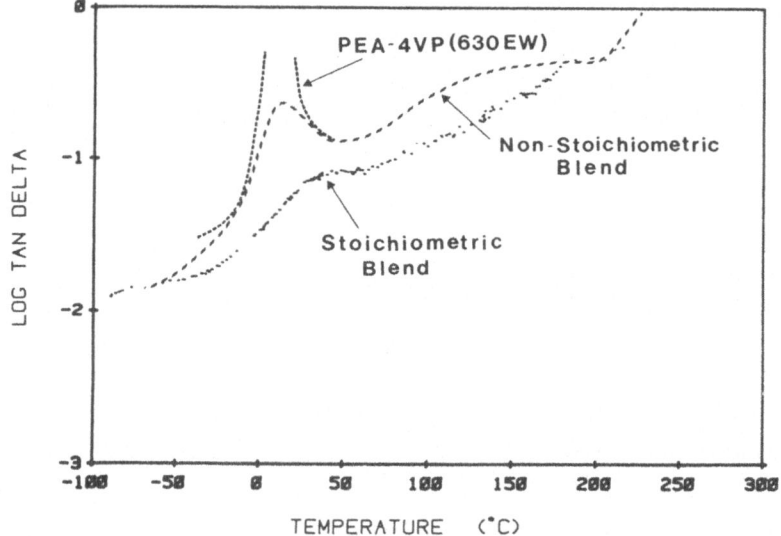

Figure 5. Variation of the loss tangent (tan δ) with
temperature for materials mentioned in Figure 4.

cussed later. These results are also reflected in the tan δ
curves of these blends. For example, a strong tan δ peak due to
the presence of the ethyl acrylate phase is observed for the
stoichiometric blends of S Ø P Ø 0 - Acid (1200 EW)/PEA-4VP
(1960 EW) as well as the S Ø P Ø 0 - Acid (1200 EW)/PEA-4VP
(952 EW). By contrast, the intensity of the tan δ peaks due to
the ethyl acrylate component in S Ø P Ø 0 - Acid (600 EW)/PEA-
4VP (630 EW) (49/51 w/w) and S Ø P Ø 0 - Acid (510 EW)/PEA-4VP
(480 EW) blends is found to be much less.

The effect of blending temperatures on miscibility enhancement is also investigated by blending the S Ø P Ø 0 - Acid (510 EW) with PEA-4VP (480 EW) at 90°C and comparing the results with samples prepared at room temperature. The dynamic mechanical properties of these blends are shown in Figure 8. It should be noted that the intensity of the tan δ peak decreases further in

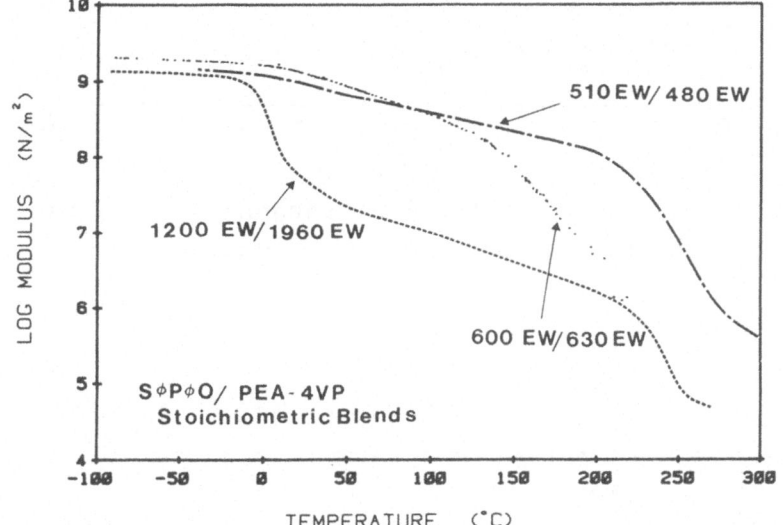

Figure 6. Variation of the storage modulus (G') with temperature for S Ø P Ø 0 - Acid (1200 EW) / PEA-4VP (1960 EW) blend 38/62 w/w), S Ø P Ø 0 - Acid (600 EW) / PEA-4VP (630 EW) blend (49/51 w/w) and S Ø P Ø 0 (510 EW) / PEA-4VP (480 EW) blend (52/48 w/w). Experimental points are only shown for S Ø P Ø 0 - Acid (600 EW) / PEA-4VP (630 EW) blend (49/51 w/w).

the case of a sample prepared at 90°C and also the G' vs T curve is smoother in that range by comparison with the sample prepared at room temperature. A non-stoichiometric blend of S Ø P Ø 0 - Acid (510 EW)/PEA-4VP (480 EW) (27/73 w/w) prepared at 90°C also exhibits similar behavior in comparison with a sample prepared at room temperature (Figure not shown).

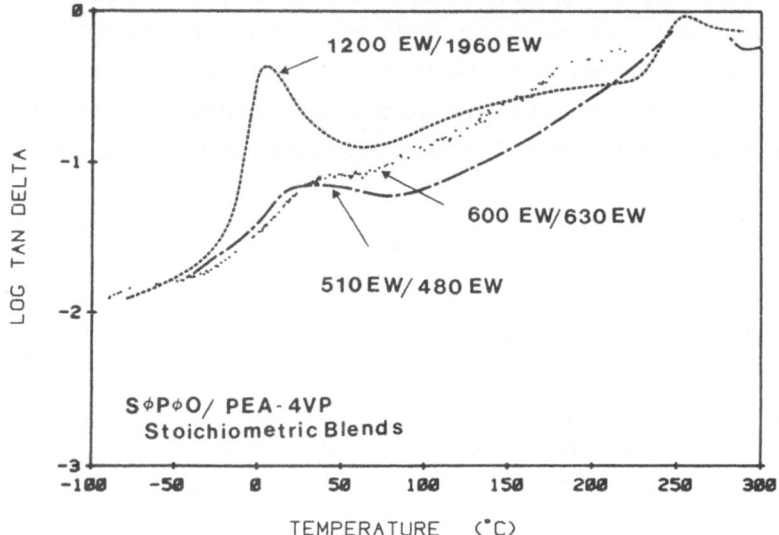

Figure 7. Variation of the loss tangent (tan δ) with temprature for blends mentioned in Figure 6.

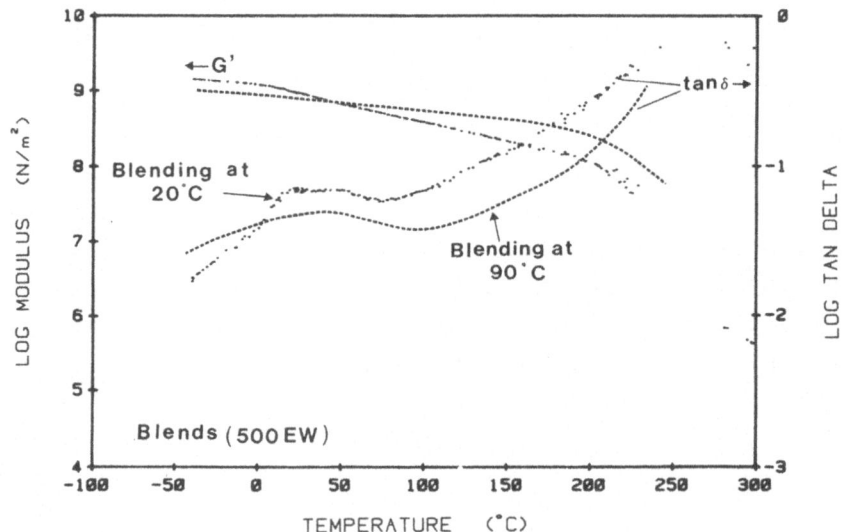

Figure 8. Variation of the storage modulus (G') and loss tangent (tan δ) with temperature for stoichiometric blends of S Ø P Ø O - Acid (510 EW) / PEA-4VP (480 EW) (52/48 w/w) prepared at room temperature and at 90°C.

To determine whether the ionic interactions are indeed responsible for the miscibility enhancement, and not the dipolar ones, blends of PEA-4VP copolymers containing varying concentrations of 4-vinyl pyridine mixed with the unfunctional-ized Ø P Ø O homopolymer were also studied. The dynamic mechanical results of blends containing the lowest and the highest 4-vinyl pyridine level are shown in Figure 9. As seen before for immiscible blends, a two-step descent in the G' curve as well as the presence of two tan δ peaks are apparent also in these samples. It should be noted that the glass transition values of the ethyl acrylate component in the blends are found to be slightly higher than the Tg's of the individual PEA-4VP copolymers (Table 2). Also, the Tg of the Ø P Ø O component in the blend is moved to a lower temperature by comparison with the Tg of the homopolymer. These results will be discussed in the next section.

To confirm independently that the ionic interactions are due to proton transfer from the acid polymer to the base polymer[16], an FTIR study was carried out on the Ø P Ø O/PEA ionic blends.

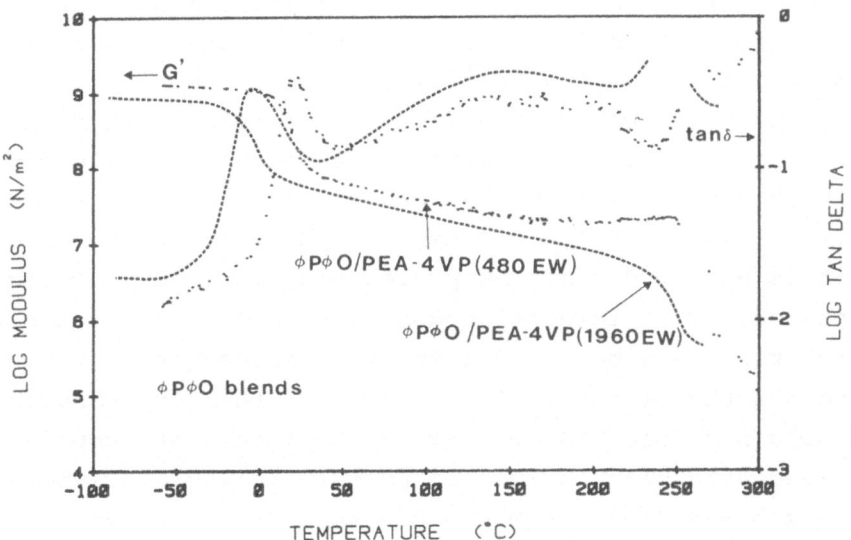

Figure 9. Variation of the storage modulus (G') and loss tangent (tan δ) with temperature for Ø P Ø O / PEA-4VP (1960 EW) (50/50 w/w) and Ø P Ø O / PEA-4VP (480 EW) (50/50 w/w).

Table 2.   Effects of PEA-4VP with varying concentrations of
4-VP's on Tg's of ethyl acrylate and Ø P Ø 0
components in the blends.

Material (50/50 w/w)	Tg of PEA-4VP in the blend (°C)	ΔTg of PEA-4VP in the blend (°C)*	Tg of Ø P Ø 0 in the blend (°C)	ΔTg of Ø P Ø 0 in the blend (°C)+
Ø P Ø 0/PEA	-3	13	245	-35
Ø P Ø 0/PEA-4VP (1960)	0	8	246	-34
Ø P Ø 0/PEA-4VP (630)	16	6	245	-35
Ø P Ø 0/PEA-4VP (510)	24	4	255	-25

*ΔTg of PEA-4VP in the blend = Tg of PEA-4VP in the blend –
Tg of pure PEA – 4VP
copolymers
+ΔTg of Ø P Ø 0 in the blend = Tg of Ø P Ø 0 in the blend –
Tg of pure Ø P Ø 0

The FTIR spectra of an immiscible blend of Ø P Ø 0/PEA-4VP
(630 EW) (50/50 w/w) as well as that of the functionalized
blend of S Ø P Ø 0 – Acid (600 EW/PEA-4VP (630 EW) (49/51 w/w)
are shown in Figure 10.   The appearance of a new absorbance
band at 1636 cm^{-1} due to the vinyl pyridinium group (VPH$^+$)
and a complete disappearance of absorbance bands due to 4-vinyl
pyridine at 1414 and 1555 cm^{-1} are evident.

The extent of proton transfer was determined semiquanti-
tively using a calibration curve of the ethyl acrylate copolymers, which had previously been neutralized completely with
perchloric acid.   Because of the high pKa value of perchloric
acid, it is expected that the protonation of 4-vinyl pyridine
is extensive.   This is confirmed by the appearance of the vinyl
pyridine absorbance band and complete disappearance of vinyl
pyridine absorbance bands. The ratio of $^{\upsilon}$VPH$^+$/$^{\upsilon}$C=0 absorbance
values obtained from the spectra of ethyl acrylate copolymers
neutralized with perchloric acid as well as for ethyl acrylate
copolymers blended with S Ø P Ø 0 – Acid are given in Table 3.
For the stoichiometric blends of S Ø P Ø 0 – Acid (600 EW)/
PEA-4VP (630 EW) and S Ø P Ø 0 – Acid (510 EW)/PEA-4VP (480
EW), the approximate percent proton transfer was found to be
between 80 and 85.   These results suggest that the functional
groups are presented mostly as ion pairs.

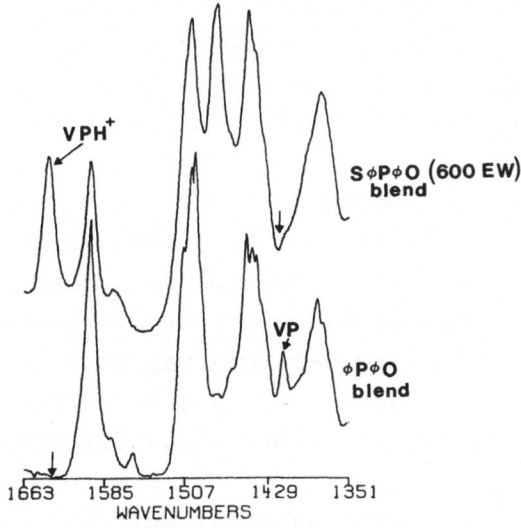

Figure 10. FTIR spectra of Ø P Ø O / PEA-4VP (630 EW) (50 / 50 w/w) and S Ø P Ø O - Acid (600 EW) / PEA-4VP (630 EW) (49/51 w/w) blend.

Table 3. FTIR results of PEA-4VP copolymers neutralized with perchloric acid as well as for PEA-4VP copolymers blended with S Ø P Ø O - Acid.

PEA-4VP copolymers neutralized with perchloric acid			PEA-4VP copolymers blended with S Ø P Ø O - Acid			
VPH mol%	EW	$\dfrac{\upsilon_{VPH}}{\upsilon_{C=0}}$	EW of S Ø P Ø O Acid	Ratio of S Ø P Ø O-Acid/ PEA-4VP (w/w)	$\dfrac{\upsilon_{VPH}}{\upsilon_{C=0}}$	approx. % proton transfer
5.1	1960	0.179	1200	38/62	0.071	65
10.5	952	0.263	1200	56/44	0.158	60
16.0	630	0.470	600	49/51	0.370	80
21.0	480	0.561	510	52/48	0.465	83

DISCUSSION

As is evident from Figures 2 and 3, the blend of unfunctionalized Ø P Ø O and PEA homopolymers is immiscible; it
exhibits a two-step descent in G' as well as two tan δ peaks.
This is not surprising in view of the very large difference in
glass transitions. However, the glass transition temperature
of the PEA component in the blend is slightly higher than the
value obtained for pure PEA, the difference being 13°C (Table
2). Furthermore, the Tg of the Ø P Ø O component in the blend
is moved to a lower temperature. The shift in the glass
transition values may be due to the presence of weak inter-
tions, e.g. of the dipole-dipole type between these polymer
pairs, which suggests that they may be slightly compatible, at
least at the interface. It may also be due to interfacial
effects arising from the very large difference in glass transitions of the pure phases.

Even if functionalized ethyl acrylate copolymers are blended with the unfunctionalized Ø P Ø O homopolymer, no improvement in miscibility is found. Furthermore, variations in the
concentrations of 4-vinyl pyridine in the copolymer also have
no significant effect on miscibility of these polymer pairs
(Figure 9). Dipolar interactions are therefore not sufficient
to induce miscibility.

Incorporation of sufficient sulfonic acid groups in Ø P Ø O
(EW ≤ 600) results in a dramatic improvement in the miscibility
as is clearly evident from the dynamic mechanical results of
the stoichiometric blends of S Ø P Ø O Acid (600 EW)/PEA-4VP
(630) and S Ø P Ø O - Acid (510 EW/PEA-4VP (480 EW) (Figures 6
and 7). These materials exhibit only a single, although broad
descent in the storage modulus curve as well as a dramatic
decrease in the intensity of the tan δ peak due to the ethyl
acrylate component in comparison with the results of the non-
ionic blends. The breadth of the drop in G' with T in the
ionic blends suggests that the mixture is heterogeneous, but
these blends do exhibit strong miscibility because of the
presence of ionic interactions resulting from proton transfer
from the sulfonic acid polymer to the 4-vinyl pyridine copolymer.

An increase in the blending temperature from room temperature to 90°C leads to considerable improvement in the miscibility. This can be seen from the shape of the G' curve in the

range of the Tg of PEA-4VP and from the decrease in the intensity of the tan δ peak due to the ethyl acrylate component. This increase in miscibility is due to increased coil overlapping, as was observed by NMR techniques in the case of PS/PMMA and Ø P Ø 0/PMMA ionic blends[21,29]. An upfield shift of the proton signals of the methoxy protons in PMMA due to the screening effect of the neighbouring aromatic rings was seen in these blends.

The specific interactions involved in these miscible blends, as mentioned before, are due to the presence of ionic interactions which originate from a proton transfer mechanism. This is clearly evident from the FTIR studies of these blends. Ionic blends show an absorbance band at 1634 cm^{-1} due to vinyl pyridinium groups as a result of proton transfer from the sulfonic acid to the 4-vinyl pyridine ring. Also, absorbance bands characteristic of 4-vinyl pyridine disappear completely for the stoichiometric S Ø P Ø 0 - Acid (510 EW)/PEA-4VP (480 EW) and S Ø P Ø 0 - Acid (600 EW)/PEA 4-VP (630 EW) blends. The extent of proton transfer in these samples is found to be between 80 and 85% (Table 3). Thus, these results suggest that considerable miscibility enhancement in these systems is indeed achieved through Coulombic interactions between the sulfonate and vinyl pyridinium groups, in spite of the very large glass transition differences.

CONCLUSIONS

Unfunctionalized Ø P Ø 0 and PEA differ in their Tg by 300°C. A blend of these two homopolymers is immiscible as seen from the opacity of the sample, a two-step descent in the G′ curve and the presence of two tan δ peaks. The incorporation of sulfonic acid groups on the Ø P Ø 0 chains and 4-vinyl pyridine on the PEA chains results in a dramatic improvement in the miscibility. In the case of stoichiometric blends of functionalized polymers of 500 to 600 EWs, a single but broad descent in G′ vs T plot and a dramatic decrease in the intensity of the tan δ peak due to the ethyl acrylate phase are seen. Blends prepared at 90°C show a further improvement in the miscibility due to enhanced coil overlap. Ionic interactions due to a proton transfer mechanism are clearly evident from FTIR studies.

Thus, considerable miscibility enhancement in Ø P Ø O/PEA blends has been achieved via ionic interactions.

ACKNOWLEDGEMENT

This work was supported in part by NSERC Canada and by U.S. Army Research Office.

REFERENCES

1. D. Klempner and K. C. Frisch, Eds., "Polymer Alloys: Blends, Blocks, Grafts and Interpenetrating Networks", Polym. Sci. and Tech., 10, Plenum Press, New York (1977)

2. D. R. Paul and S. Newman, Eds., "Polymer Blends", 1 and 2, Academic Press, New York (1978)

3. O. Olabisi, L. M. Robeson and M. T. Shaw, "Polymer - Polymer Miscibility", Academic Press, New York (1979)

4. S. L. Cooper and G. M. Estes, Eds., "Multiphase Polymers", ACS Adv. Chem. Ser., No. 176, ACS, Washington, D. C. (1979)

5. L. H. Sperling, "Interpenetrating Polymer Networks and Related Materials", Plenum Press, New York (1981)

6. C. D. Han, Ed., "Polymer Blends and Composites in Multiphase Systems", ACS Adv. Chem. Ser., No. 206, ACS, Washington, D. C. (1984)

7. D. R. Paul and L. H. Sperling, Eds., "Multicomponent Polymer Materials", ACS Adv. Chem. Ser., No. 211, ACS, Washington, D. C. (1986)

8. L. M. Robeson, Polym. Eng. and Sci., 24, 587 (1984)

9. R. E. Prud'homme, Polym. Eng. and Sci., 22, 90 (1982)

10. R. Longworth, U.S. Pat. 3,299, 176 (1967)

11. S. Djadoun, R. N. Goldberg, and H. Morawetz, Macromolecules, 10, 1015 (1977)

12. E. M. Pearce, T. K. Kwei and B. Y. Min, J. Macromol. Sci., Chem., A21, 1181 (1984)

13. T. Sulzberg and R. J. Cotter, J. Polym. Sci., Part A-1, 8, 2717 (1970)

14. C. Pugh and V. Percec, Macromolecules, 19, 65 (1986)

15. M. Hara and A. Eisenberg, Macromolecules, 17, 1335 (1984)

16. P. Smith and A. Eisenberg, J. Polym. Sci., Polym. Lett., <u>21</u>, 223 (1983)

17. Z. L. Zhou and A. Eisenberg, J. Polym. Sci., Polym. Phys., <u>21</u>, 595 (1983)

18. M. Rutkowska and A. Eisenberg, Macromolecules, <u>17</u>, 821 (1984)

19. D. G. Peiffer, I. Duvdevani, P. R. Agarwal and R. D. Lundberg, J. Polym. Sci., Polym. Lett., <u>24</u>, 581 (1986)

20. C. Beretta and R. A. Weiss, Proceedings of the ACS Div. of Polym. Materials Sci. and Eng., <u>56</u>, 556 (1987)

21. A. Natansohn and A. Eisenberg, Macromolecules, <u>20</u>, 323 (1987)

22. E. P. Cizek, U.S. Pat., 3,383,435 (1968)

23. W. J. MacKnight, J. Stoelting and F. E. Karasz, Polym. Eng. Sci., <u>10</u>, 133 (1970)

24. F. E. Karasz, W. J. MacKnight and J. Stoelting, J. Appl. Phys. <u>41</u>, 4357 (1970)

25. D. Duchesne, Ph. D. Thesis, McGill University (1985)

26. J. E. Burleigh, O. F. McKinney and M. G. Barker, Analytical Chem., <u>31</u>, 1684 (1959)

27. H. S. Makowski, R. D. Lundberg and G. H. Singhal, U.S. Pat. 3,870,841 (1975)

28. R. Murali, A. Eisenberg, R. K. Gupta and F. W. Harris (to be submitted)

29. R. Murali, A. Natansohn, R. Tannenhaum and A. Eisenberg (to be submitted)

CONFORMATIONAL TRANSITIONS AS MOLECULAR SIGNALLING MECHANISMS IN SYNTHETIC BILAYER MEMBRANES

David A. Tirrell[a], Kenneth H. Langley[b], Ki Min Eum[b] and Keith A. Borden[a]

Department of Polymer Science and Engineering[a] and Department of Physics and Astronomy[b], University of Massachusetts, Amherst, MA 01003

Conformational transitions in membrane-bound macromolecules probably provide the most general and powerful mechanisms for chemical and physical signalling processes in biology. Changes in the conformations of membrane proteins and glycoproteins are exploited in cell biology to modulate enzymatic activity, to change membrane permeability, and to initiate the membrane reorganization events involved in endocytosis, in secretion and in the processing of ligands and receptors. In recent years, we have explored the consequences of conformational transitions in *synthetic* macromolecules bound to lipid bilayer membranes by adsorption or via hydrophobic anchoring groups.[1-4] The present chapter examines the roles of polymer conformation and solvation in controlling membrane structure in such systems.

The pH-dependent solution properties of poly(2-ethylacrylic acid) (PEAA, **1**) have been examined by Fichtner and Schonert,[5] by Joyce and Kurucsev[6] and by Sugai and coworkers.[7] The potentiometric titration behavior of PEAA suggests that the polymer undergoes a coopera-

$$\begin{array}{c} CH_2CH_3 \\ | \\ +CH_2C\!+_n \qquad\qquad \textbf{1} \\ | \\ CO_2H \end{array}$$

tive conformational transition from an expanded coil at high pH to a compact globule upon acidification. Figure 1 shows the results of fluorescence and light scattering experiments that are consistent with this view. The Figure reports: i). the pH-dependence of the pyrene emission intensity (I_f) at 375 nm for solutions of 5 μM pyrene is Tris-buffered aqueous PEAA, and ii). the hydrodynamic radius (R_H) of PEAA dissolved in similar buffers.[8,9] The sharp decrease in R_H upon acidification from pH 6.2 to pH 5.9 is nearly coincident with a ca. six-fold rise in I_f. R_H changes from ca. 6 nm at high pH to ca. 4.4 nm in acidic solutions, and the rise in I_f signals the conformational collapse of the polymer to a globular coil in which pyrene is solubilized and pyrene-water contacts are reduced.[10]

Figure 2 illustrates that similar information can be obtained from the pH-dependent fluorescence of a *polymer-bound* pyrene probe.[4] The conformational transition of a copolymer of 2-ethylacrylic acid and 1-pyreneacrylic acid appears in the fluorescence experiment as a large increment in I_f around pH 6.2. Covalent binding of the chromophore has little effect on the position or the shape of the transition.

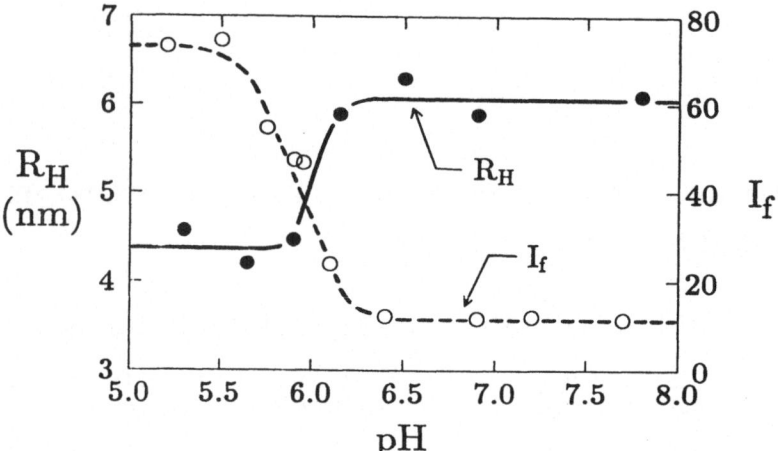

Figure 1. Conformational transition of PEAA upon pH variation. (●) Hydrodynamic radius of PEAA (R_H); 0.5 mg/ml PEAA in 50n mM Tris-HCl buffer (pH 7.8); pH varied by 1 N HCl. (○) Fluorescence intensity (I_f) (arbitrary units); PEAA 1 mg/ml + pyrene 5 μM in 50 mM Tris-HCl buffer.

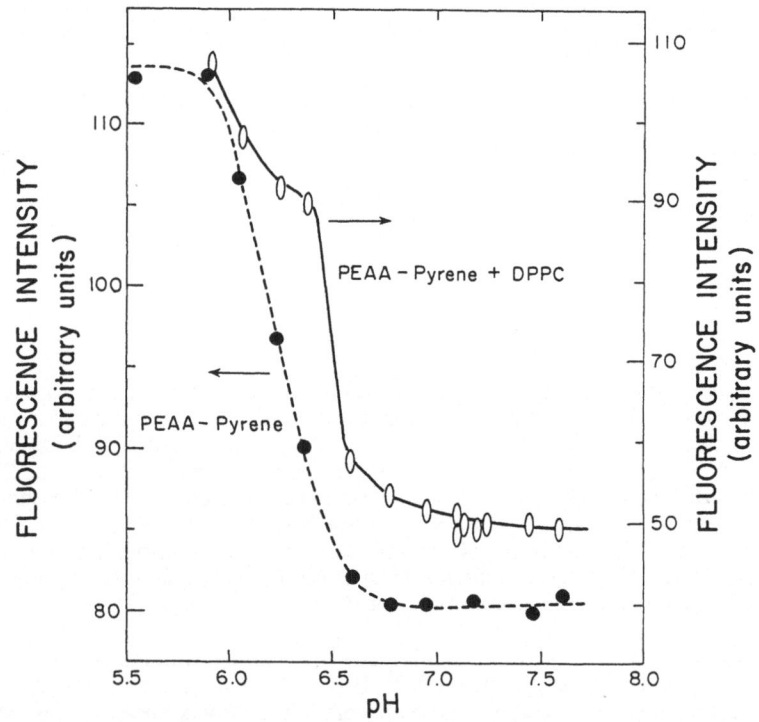

Figure 2. Fluorescence intensity (379 nm) emitted by 2-ethylacrylic acid/1-pyreneacrylic acid copolymer in phosphate buffers of differing pH. (●) Lipid-free polymer solutions; (0) Polymer (0.9 mg/mL) plus DPPC (1.0 mg/mL).

58

Also shown in Figure 2 is the behavior of the copolymer when dissolved in buffered suspensions of dipalmitoylphosphatidylcholine (DPPC, **2**). The transition midpoint is shifted from pH 6.2 to pH 6.5 by the addition of an equivalent weight of the lipid.

$$CH_2OPOCH_2CH_2N^+(CH_3)_3$$

with structure showing:

$$
\begin{array}{c}
\quad\quad O \\
\quad\quad \| \\
CH_2OPOCH_2CH_2N^+(CH_3)_3 \\
| \quad\quad\quad | \\
| \quad\quad\quad O^- \\
| \\
CHOCOC_{15}H_{31} \\
| \\
CH_2OCOC_{15}H_{31}
\end{array}
\quad\quad\quad\quad \mathbf{2}
$$

Figure 3 demonstrates that *these changes in solvation of the polymer-bound chromophore are coincident with membrane reorganization.* The Figure re-plots the fluorescence data for the copolymer, and shows that the increase in I_f may be superimposed on an increase in the calorimetric phase transition width ($\Delta T_{1/2}$) determined for the mixed polymer-lipid aggregate. Thus the pH-dependent change in lipid organization that broadens the melting transition is accompanied by an abrupt change in hydration of the polymer chain. The fact that these changes occur at higher pH than collapse of the free chain (ca. pH 6.5 vs. pH 6.2) argues that adsorption must precede conformational collapse in a pseudoequilibrium experiment in which acidification is infinitely slow (Figure 4). Thus as the pH is depressed, soluble PEAA is adsorbed in increasing amounts at the membrane surface. Further acidification (and concomitant depression of chain ionization) causes desolvation of the adsorbed chains and reorganization of the aggregate to a mixed micelle in which water-polymer (and water-pyrene) contacts are reduced. An alternative description of this process, in which soluble PEAA first collapses and then partitions into the bilayer (Figure 4, route a) is inconsistent with these results.

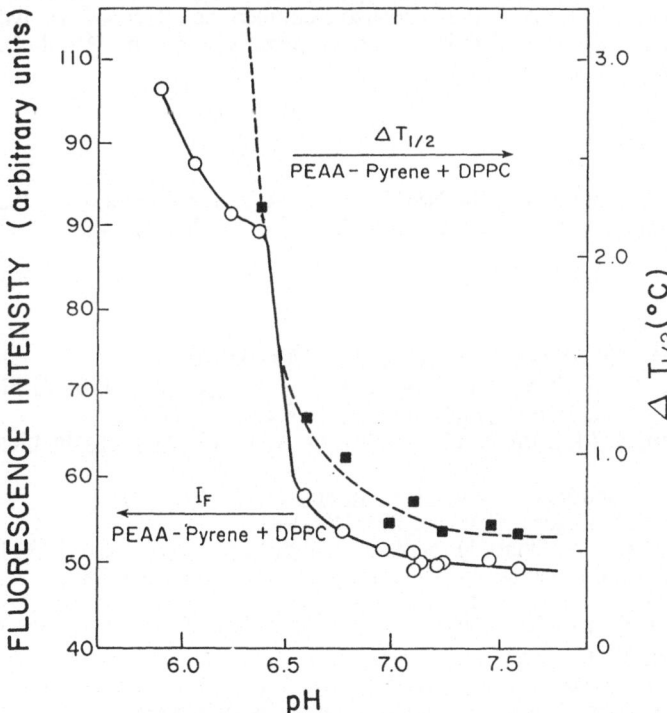

Figure 3.   Fluorescence intensity emitted by PEAA/DPPC mixtures as in Figure 2 (○) and calorimetric transition half-width $\Delta T_{1/2}$ (■) for the same samples. $\Delta T_{1/2}$ at pH 6.25 is 5.5°C (not shown).

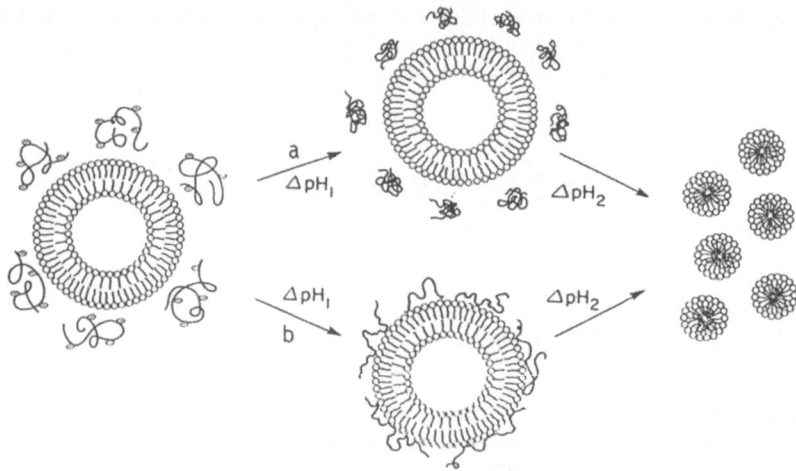

Figure 4.   Schematic illustration of the PEAA-driven vesicle-to-micelle transition in DPPC. Route a (upper): Acidification causes conformational collapse of PEAA and subsequent partitioning of the collapsed chain into the bilayer. Route b (lower): Acidification causes polymer adsorption without conformational collapse. Further protonation causes desolvation of PEAA and reorganization of the aggregate into mixed micelles. The latter route appears to describe the pseudoequilibrium experiments discussed in the text. The sequence of events that follows *rapid* acidification has not been determined.

We have exploited these changes in polymer conformation and solvation to prepare phosphatidylcholine vesicles with permeability properties subject to modulation by pH,[1,2] temperature,[2] glucose[3] or light.[11] We have also examined the effects of: i). hydrophobic anchoring of PEAA,[12] ii). the molecular weight of the polyelectrolyte,[13] and iii). the composition of the lipid bilayer.[8,9]

ACKNOWLEDGMENTS

This work was supported by the NSF Materials Research Laboratory at the University of Massachusetts and by a grant from the U.S. Army Research Office (DAAL03-88-K-0038).

REFERENCES

1. K. Seki, D. A. Tirrell, Macromolecules 17:1692 (1984).
2. D. A. Tirrell, D. Y. Takigawa, K. Seki, Ann. N.Y. Acad. Sci. 446:237 (1985).
3. B. P. Devlin, D. A. Tirrell, Macromolecules 19:2465 (1986).
4. K. A. Borden, K. M. Eum, K. H. Langley, D. A. Tirrell, Macromolecules 20:454 (1987).
5. F. Fichtner, H. Schönert, Colloid Polym. Sci. 255:230 (1977).
6. D. E. Joyce, T. Kuruscev, Polymer 22:415 (1981).
7. S. Sugai, K. Nitta, N. Ohno, H. Nakano, Colloid Polym. Sci. 261:159 (1983).
8. K. M. Eum, Ph.D. Dissertation, University of Massachusetts (1988).
9. K. M. Eum, K. H. Langley, D. A. Tirrell, Macromolecules, submitted.
10. T. S. Chen, J. K. Thomas, J. Polym. Sci., Polym. Chem. Ed. 17:1103 (1979).
11. M. S. Ferritto, D. A. Tirrell, Macromolecules, in press.
12. M. Maeda, A. Kumano, D. A. Tirrell, J. Am. Chem. Soc., in press.
13. U. K. O. Schröder, D. A. Tirrell, Macromolecules, submitted.

ELECTROCHEMICALLY CONTROLLED RELEASE

OF IONS FROM POLYMERS

Larry L. Miller

Department of Chemistry
University of Minnesota
Minneapolis, MN 55455

INTRODUCTION

This work has been motivated by an interest in the application of organic chemistry and electrochemistry to biomedical science. It has grown from studies performed at Colorado State University and now the University of Minnesota on chemically modified electrodes. Following our initial discovery that electrode surfaces could be covalently modified, we turned some eleven years ago to modification of electrodes with polymers.[1] This approach, which generally involves coating a metal or semiconductor electrode with a thin polymer film, has proven widely interesting to electrochemists.[2] They have studied a variety of electrochemical phenomena which result when redox polymers are coated on electrodes. Applications to catalysis, solar energy conversion and sensors (among others) have been explored. Our interest over the past several years has been in using these polymer coated electrodes as devices to bind and release organic compounds, especially compounds of biomedical importance.

Two general approaches have been explored. In one, the major topic of this paper, ionic binding was used. In the other, the compound of interest was covalently bound to a polymer by a cathodically cleavable bond. This polymer was constructed with an "electrophore" present so that the cathodic reaction would be possible at a not–too–negative potential. In practice, the following polymer on a carbon electrode gave release of dopamine into aqueous solution upon cathodic reduction.[3]

This approach has the advantage that the compound of interest cannot leak out, but it has several disadvantages. It is, in particular, difficult to prepare polymers which have the requisite properties so that prompt and high yield release of the compound from thick films is feasible. The above polymer proved deficient in that charge did not readily propagate through the film and only tiny amounts could be released.

Charge propagation through redox polymer films has become a subject of great interest to electrochemists in this field. The process requires that electrons are transported from one redox site and polymer molecule to another. In addition, there must be ion transport to maintain electroneutrality, and morphological changes to accommodate the change in chemical composition. Typically the combined processes of electron and ion diffusion can be characterized by some apparent diffusion coefficient.

We realized that good charge propagation would be possible using conducting polymers[4] like polypyrrole and report here several examples in which ions, especially organic ions, have been ionically bound into heterocyclic conducting polymers and released electrochemically.

The redox properties of conducting heterocyclic polymers like polypyrrole are central to many applications of these materials. For this reason the electrochemistry of thin films of these polymers have received a lot of attention. For use in electrically controlled ion binding and delivery, the general idea is that the heterocyclic conducting polymers have cationic backbones and will incorporate counter anions. Upon reduction of the backbone, the anions will be flushed out. Thus, in principle one can develop devices to absorb anions of interest or to release them in response to an electric current. Our work has been spurred by the possibility of delivering drugs with a rate controlled by the current.

New methods of drug delivery are of intense interest and transdermal controlled release of drugs is a field of importance.[5] Controlled release typically aims for and achieves a zero–order release rate. This ensures that the drug concentrations in vivo remain in an appropriate range over an extended period of time. Our interest is in a more sophisticated system, where the release could be turned on and off, and varied in some appropriate way. It might, for example, be used to respond to metabolic changes. One approach to this problem is to use electrochemistry to drive ionic drugs through the skin. The stratum corneum prevents transport of most materials through the skin, especially ions, but an electrochemical potential can be used to force ions through. This method is called transdermal iontophoresis.[5]

Results from in vitro experiments in which ^{14}C–labelled acetate was iontophoresed from an aqueous solution through excised hairless mouse skin are shown in Figure 1.[6] Using a constant current, a constant flux of acetate through the skin is achieved. The effects on the flux (J_A) of changing the current (I) and the concentration of sodium acetate in the donor solution are represented in the figure.

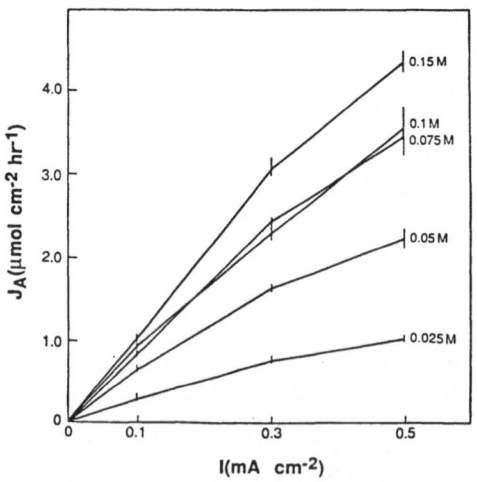

Fig. 1.   Dependence of J_A on I and acetate concentration
in the donor solution.

A conducting polymer would be used in a transdermal iontophoretic device as the reservoir for drug ions. The ions would be metered out by controlling the number of coulombs passed through the electrical circuit which involves (at least) an electrode coated with polymer, the skin, the body fluid, the skin, and a second electrode placed elsewhere on the skin. Drug ions are forced out of the polymer and through the skin. Other ions (Na^+, Cl^-) carry the current in the body and back through the skin.

BINDING AND RELEASE OF ANIONS

We have reported on the binding and release of $Fe(CN)_6^{3-}$ $^{4-}$ from polypyrrole.[7] The anodic polymerization of pyrrole onto glassy carbon in the presence of aqueous $K_4Fe(CN)_6$ gave a polypyrrole which incorporated $Fe(CN)_6^{3-}$ (PP/FCN). The polymerization process for PP/FCN was studied using chronoamperometry and showed nucleation and growth kinetics. The concentration of electrolyte and pyrrole governed the nucleation and growth rates. It was shown that stirring inhibited both nucleation and growth, suggesting the involvement of soluble species in film formation. In these PP/FCN films, the FCN did not readily exchange with other ions in aqueous solution and could only be flushed out of a film formed with $Q = 2$ mC cm^{-2} into 0.1 M NaCl at –0.4 V (SCE). It was shown that the amount of FCN$^-$ released depended on the film thickness. It was shown that little FCN is cathodically released into aqueous CsCl or tetraethylammonium chloride solutions. X–ray photoelectron spectroscopy indicated instead that Cs$^+$ is bound into the reduced PP/FCN film. It was demonstrated that repetitive 5–s pulses to –0.4 V from open circuit gave controlled release of FCN into a small volume of solution. PP/FCN films changed, giving smaller release amounts, when they were allowed to dry.

These results, which revealed the competition between cathodic anion release and cathodic cation, e.g. Cs$^+$, binding into PP$^+$ FCN$^-$, led us to a further examination of this phenomenon using PP$^+$ ClO$_4^-$ and PP$^+$ OTs$^-$.[8] The results demonstrated that the nature of the anion caused major differences in the competition between anion loss and cation gain. Perchlorate, for example, was more easily lost than tosylate into aqueous NaCl solution. It is hypothesized that the more hydrophobic tosylate is more strongly bound and preferentially retained.

Using a film of anodically–prepared, oligomeric 3–methoxythiophene, we have studied the binding and release of several anions.[9] Soluble, oligomerized 3–methoxythiophene was was purified and deposited from acetonitrile solution onto carbon electrodes. Upon oxidation at 0.4–0.6 V (SCE) in aqueous solutions of the anions salicylate or ferricyanide, the film was oxidized and incorporated the anions to maintain electroneutrality. These anion–loaded films on carbon were placed in aqueous NaCl and then the anions were released by stepping the electrode potential to pass cathodic current, e.g. to –0.4 V. In the case of salicylate, detection of the released anion by emission spectroscopy showed that at open circuit much of the salicylate exchanged out of the film over a period of ten minutes, but release of the salicylate could be completed in only thirty seconds by stepping the potential to –0.4 V. There was a linear relationship between coulombs passed and the amount of salicylate released. Ferricyanide release into aq. NaCl was monitored electrochemically. Exchange of ferricyanide for chloride at open circuit was slow on the time scale of minutes, but release was completed in 30 sec at –0.2 V.

BINDING AND RELEASE OF ORGANIC CATIONS

Since suitable conducting polymers with anionic backbones are not available, we have resorted to a more complex approach for cation binding. Following Martin's work[10] on terpolymer redox films of polyvinylferrocene, we produced a new type of composite polymer, of cationic poly–(N–methylpyrrole) (PMP$^+$) with anionic polystyrenesulfonate (PSS$^-$). Upon reduction of a film of this material, cations are taken up. Thus, the entangled polymeric anion is not flushed out by reduction of the pyrrole units. Instead, cations are incorporated to balance the sulfonate charges. This has been shown for a variety of cations including protonated dopamine. The scheme below shows how this polymer works to bind protonated dimethyldopamine (DH$^+$) cathodically and to release it anodically.

$$e^- + PMP^+ PSS^- + \underset{DH^+}{\overset{\displaystyle CH_2CH_2\overset{+}{N}H_3}{\underset{\displaystyle CH_3O \quad\quad OCH_3}{\bigcirc}}} \rightleftharpoons PMP^\circ PSS^- DH^+$$

eqn. 1

The general method of anodic polymerization has been reported. The PMP$^+$/PSS$^-$ composite film was prepared by electrolysis of 0.05 M N–methylpyrrole, and 0.03–0.1 M aqueous PSS–Na solutions at a constant potential, e.g., +0.7 V vs. SCE, or by a constant current, e.g., 2 – 4 mA/cm^2. The resulting film was shiny dark brown when it was thin and dull black when it was thick. The color of the reduced film varied from golden–yellow to brown–yellow as the film thickness increased. For the electropolymerization at constant potential, the current–time transients showed that the electrolysis current continuously declined. It was found that this tendency was more severe for a higher concentration of PSS–Na solution and very dependent on stirring. Electropolymerization with a constant current was much more satisfactory. For example, it needed stirring, but not a critical stirring speed. The electrode potential was rather stable and reproducible during these constant current experiments.

The film thickness and capacity were determined for a set of PMP$^+$/PSS$^-$ films. The thickness (δ) was estimated using a profilometer and compared to the coulombs passed in film formation. There was a linear Q/δ relationship showing 26.2 mC cm^{-2} produces a 0.1 μm film. This value is quite close to the literature value for PP$^+$/ClO$_4^-$. The characterization of this composite polymer as well as its reduced (in aq NaCl) partner PMP$^\circ$/PSS$^-$/Na$^+$ using ESCA, elem. anal., EM, and IR has been reported.[11]

Dopamine Release

Initial studies showed that protonated dopamine could be bound into PMP$^+$/PSS$^-$ by reducing the film at −0.5 V (SCE) in aqueous dopamine hydrobromide. This loaded film was then placed in aqueous NaCl, and the release of dopamine was detected electrochemically on a second electrode. Release was slow unless an anodic step to 0.5 V was applied. Such a step gave prompt, i.e. within minutes, release.

Initial spectroscopic experiments[12] utilized dopamine and a simple spectroelectrochemical cell. The UV light beam, which was wider than the gap between electrodes, passed through the solution separating the PMP$^+$/PSS$^-$ coated Pt electrode and the counter, Pt, electrode. Since it was shown that the dopamine in the film did not absorb and the dopamine in solution followed Beer's Law, this set–up provided a method to quantitate the total amount of dopamine in solution. Since the release was done in the light beam, the time course of release could be examined. The data made it clear that there was some loss of dopamine from the film into solution at open circuit, and that anodic current from a potential step to 0.4–0.6 V stimulated release.

Three PSS–Na samples with different molecular weights and different substitution percentages were employed to synthesize polymer composites with $\delta = 2.8$ μm. These were subsequently loaded with dopamine at −0.6 V. After being thoroughly rinsed, the loaded film was assembled into a UV cell for the release experiment. The dopamine absorbance–time curves indicated that the PMP$^+$/PSS$^-$ films with 100% substituted PSS$^-$ had higher capacities for dopamine than the one with 50% substituted PSS$^-$. The "leakage" into water at open circuit occurs concomitantly with a drift of the electrode potential to more positive values. Comparatively the results are consistent with electrochemical studies that showed leakage in 0.01 M aqueous LiClO$_4$ solution was faster for the film with 100% substituted PSS$^-$ than for that with 50% substituted PSS$^-$. The film with higher MW PSS$^-$ leaked into the electrolyte solution more slowly than did that with lower MW PSS$^-$.

A series of dopamine release experiments with PMP$^+$/PSS$^-$ films of different thicknesses was undertaken. The results showed that the release amount at 0.6 V increased as the film got thicker. For example, at 0.6 V 35 nmol cm^{-2} came from a 0.7 μm film, 84 nmol cm^{-2} from 2.8 μm, and 126 nmol cm^{-2} from a 3.9 μm film. Comparison of the UV spectra of the solution revealed the occurrence of a side–reaction which stems from the oxidizability of dopamine either by the air or the anodic event. Needless to say, the side–reaction interfered in the precise measurement of the dopamine released. To minimize the side–reaction, further experiments used dimethyldopamine as the drug prototype.

Dimethyldopamine Release

The general procedure was to prepare the PMP$^+$/PSS$^-$ film using 100% sulfonated, 500,000 PSS$^-$ by constant current oxidation at 2–4 mA cm^{-2}. After preparation, the coated electrode was loaded with protonated dimethyldopamine (DH$^+$) by reduction of the film at –0.6 V in 0.1 M D$\cdot$HBr. The potential was then switched to open circuit, the PMP$^\circ$/PSS$^-$/DH$^+$ electrode was rinsed with water and used immediately without letting it dry. The release was performed using 0.01 M LiClO$_4$ solution.

Using the spectroelectrochemical cell, three identically prepared 1.3 μm PMP$^+$/PSS$^-$/DH$^+$ films were soaked at open circuit in water for 20 min. and then in 0.01 M LiClO$_4$ for 20 min. Then in fresh 0.01 M LiClO$_4$, they were stepped to 0.4, 0.5 or 0.6 V. Qualitatively it is clear that the released amount as well as the charge increase over this range. Assuming that the charge was all consumed for the reoxidation of the methylpyrrole units, at 0.4 V ca. 9% of the units were shifted to the cationic state. When stepped to 0.5 V, the charge amount did not completely level off, but ca. 44% of the methylpyrrole units were oxidized to cations. With a potential of +0.6 V, the charge amount constantly increased. After 20 min, the charge amount was even more than that expected for the oxidation of all the pyrrole units. This demonstrated that a side–reaction occurs at potentials positive of 0.4 V.

Consistent with this conclusion are the efficiencies of release (moles DH$^+$/moles e$^-$). At 0.4 V, 0.05 coul produced 0.27 μmol for an efficiency of 57%. At 0.5 V the efficiency was only 15%.

Next a series of PMP$^+$/PSS$^-$/DH$^+$ films with thicknesses from 0.35 to 2.8 μ were prepared, soaked in water and oxidized by a step to 0.4 V. The final release amounts showed a linear relationship with the charge amount consumed during the pulse. This linear relationship implied that the release was indeed electrochemically initiated by the redox process. It also showed that the process was 30% efficient. For the thinner films ($\delta < 2 \mu$m), the released amount of DH$^+$ was proportional to the film thickness. This relationship indicated that the capacity of the film was proportional to the actual volume of the film. More precisely, it depended on the amounts of the oxidizable methylpyrrole units at the specific pulse potential and the amount of the polymeric counter anion which had electrostatically bound DH$^+$. All the above results suggested that the polymeric counter anion which permanently resides in the conducting polymer is responsible for binding of the cationic drug, while the redox switchable PMP can replace the drug cation in an electrochemical event.

Controlled Release of DH$^+$ by Short Multiple Pulses

Electrochemical control over release has some presumed advantages concerning the time course of release, but this had been demonstrated for the potential–step, pulsed cathodic release of Fe(CN)$_6^{4-}$ from polypyrrole. In that case there was no relationship between the coulometry and the amount released, and it was shown that the process was complicated by various factors including the instability of reduced polypyrrole.

For pulsed release of DH⁺ a 2.9 μm film was loaded and soaked in water and 0.01 M LiClO₄ as usual. It was then, in the same 0.01 M LiClO₄ solutions, pulsed to 0.6 V from open circuit for 10 sec every 5 min. As expected, the amount of DH⁺ in solution increased with time. As time progressed, the film became depleted of DH⁺, and smaller amounts were released from the 10 sec pulse. Examination of the Abs/time trace shows that the concentration goes up after the pulse and then declines at open circuit. This is do to reabsorption of DH⁺ into the film, when the potential is not held at the anodic potential.

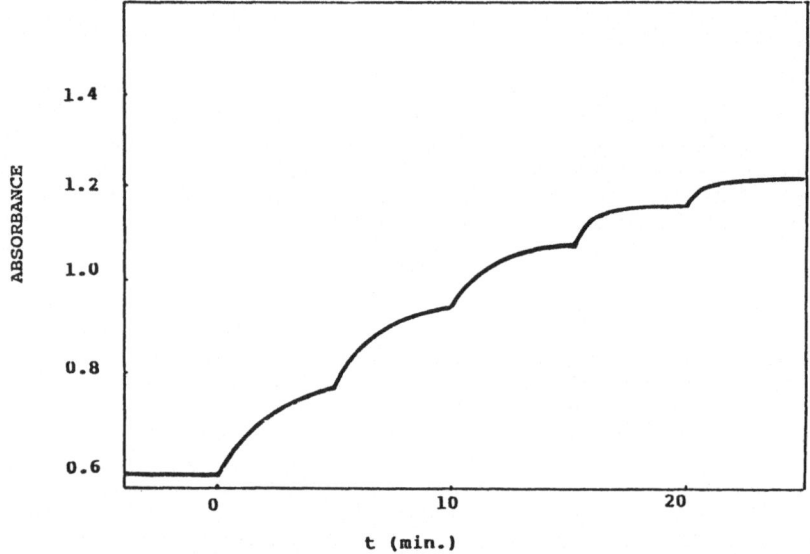

Fig. 2 Release of dimethyldopamine (DH⁺) from PMP⁺/PSS⁻ into aq. LiClO₄ in response to a 10 sec pulse to 0.4 V every 5 min.

Kinetics of Dimethyldopamine Uptake and Release

The UV method directly provided kinetic data like those in Figure 2. Such data allowed for the first time a quantitative evaluation of ion transport kinetics using a direct measure of the ion's concentration in solution. When such data were plotted as A vs $t^{\frac{1}{2}}$, the plots were linear (corr. coeff > 0.99). It is thus concluded that a diffusion process controls the rate. It was found that the rate constant (or apparent diffusion coefficient) was independent of stirring, cell dimensions, the counter electrode, and the presence of KCl. It depended on the composition of the film and, therefore, potential.

All of these data are consistent with one hypothesis: that it is diffusion in the film that determines the rate. It must be true that the motion of the electrons, the DH⁺, and the polymer chains takes place to accomplish this process; and to some extent they must be coupled.

Equilibrium Data

The above observations and others strongly suggested that when a PMP⁺/PSS⁻/DH⁺ electrode is placed in solution an equilibrium is established which corresponds to eqn. 1. It was of interest to study this equilibrium process as the potential and conditions were changed. Polyppyrroles and similar conducting polymers have generally not been well characterized in thermodynamic terms. This contrasts with polymer films containing simple redox centers like polyvinylferrocene. The primary reason for this paucity of information is that the electrochemistry, e.g. voltammetry, does not give readily analyzable data. This results in part because a lot of current flows that is unrelated to simple ion binding.

In this regard we thought it was of interest to see if we could evaluate the extent of oxidation of the PMP^+/PSS^- film spectrophotometrically. Since one DH^+ ion should be bound per electron, it seemed possible that this would reveal the charging/ion binding process independently from other competing reactions.

For this part of the study, the protocol involved preconditioning the film and then placing it into a solution containing DH^+, Br^- as the only electrolyte. Using UV, the uptake and release was then monitored over time. A sequence of potential steps was applied to the film. Typical raw data are shown in Figure 2. It can be seen that the change in $[DH^+]$ levels off after a minute or so. When the potential excursion is not too large, the same Abs. will be measured after either a positive or negative potential step. These data are sufficient to conclude that the system reaches equilibrium.

The equilibrium data (Figure 3) can be plotted in Nernst fashion assuming that the amount of DH^+ in the film corresponds to the number of PMP/PSS sites. The Nernst plot is linear with a slope of 220 mV. As expected, changing the solution concentration, and film thickness had no effect on the slope. Added 0.1 M KCl also had no effect on the slope. The slope is clearly incompatible with the concept of a single type of site isolated from other sites which would give 59 mV. The data are compatible with a band structure or with interacting sites as described by Laviron. In this regard it is worth noting that the cyclic voltammograms of PMP/PSS reveal nothing about this equilibrium situation.

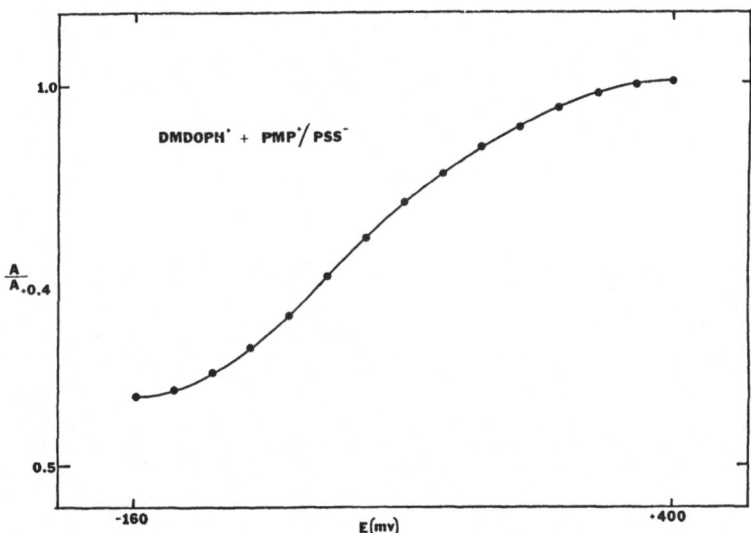

Fig. 3 Change in dimethyldopamine concentration in solution as PMP^+/PSS^- potential (SCE) is varied.

Acknowledgement: Their work was supported by the National Science Foundation and the National Institutes of Health.

REFERENCES

1. L.L. Miller and M.R. Van de Mark, *J. Am. Chem. Soc.*, **1978**, *100*, 639.
2. R.W. Murray in "Electroanalytical Chemistry" Vol. 13, A.J. Bard, Ed. 1981.
3. A.N.K. Lau and L.L. Miller, *J. Am. Chem. Soc.*, **1983**, *105*, 5271.
4. For an extensive current review see, "Handbook of Conducting Polymers," Vol I, II; T.A. Skotheim, Ed., M. Dekker, New York, 1986.
5. *Transdermal Delivery of Drugs*, A.F. Kydonieus and B. Berner Ed., CRC Press, 1987.
6. L.L. Miller and G.A. Smith, *Int. J. Pharmeceut.*, in press.
7. B. Zinger, Q.X. Zhou, and L.L. Miller, *J. Am. Chem. Soc.*, **1987**, *109*, 2267.
8. Q.X. Zhou, C.J. Kolaskie and L.L. Miller, *J. Electroanal. Chem.*, **1987**, *223*, 283.
9. (a) A.C. Chang and L.L. Miller *J. Electroanal. Chem.*, in press.
   (b) A.C. Chang, R.L. Blankespoor and L.L. Miller *J. Electroanal. Chem.*, **1987**, *236*, 239.
10. M.W. Espenscheid and C.R. Martin, *J. Electroanal. Chem.* **1985**, *188*, 73.
11. L.L. Miller and Q.X. Zhou, *Macromolecules*, **1987**, *20*, 1594.
12. Unpublished work of Q.X. Zhou and J.R. Valentine

WATER TRANSPORT IN IONIC POLYMERS

Charles E. Reineke, David J. Moll, Damoder Reddy, and
Ritchie A. Wessling

The Dow Chemical Company
Midland, Michigan 48674

ABSTRACT

Hydrophilic polymers such as perfluorinated ionomers with sulfonic
acid functional groups have high water permeabilities and good selectivities for water over most gases and organic liquids. Understanding the
water transport mechanism in these water plasticized membranes is necessary
in order to effectively utilize them for dehydration applications. Water
diffusion in these polymers is non-Fickian. Water permeabilities and
solubility coefficients are a function of water vapor pressure. A careful
analysis is required to arrive at the concentration dependent diffusion
coefficients. An approach for calculating thermodynamic diffusion
coefficients based on steady state permeabilities and equilibrium sorption
measurements is presented in this paper. At very low water vapor
pressures, strong ion-water interactions give rise to high solubility
coefficients and low diffusion coefficients. At water activities of 0.1 to
0.6, the solubility coefficient decreases and the diffusion coefficient
increases with water activity. At very high water activities ($\geq$0.6), the
diffusion coefficient appears to attain a limiting value while the
solubility coefficient sharply increases. An anomalous thickness effect
was also observed. Thickness normalized permeabilities increased with
increasing membrane thickness. Several mechanisms were considered.
Interfacial resistances which prevent attainment of the equilibrium sorption concentration in the membrane in the steady state appear to be the
plausible cause for this thickness effect.

INTRODUCTION

Copolymers of tetrafluoroethylene and sulfonic acid functional perfluorinated monomers (e.g., Nafion, Dow's perfluorosulfonic acid (PFSA))
have high water permeability. Water transport through these ionomer
membranes has been investigated. The non-Fickian diffusion process is
analyzed by a thermodynamic approach. The results provide some useful
insights into the behavior of these materials as dehydration membranes.

Dehydration of gases and liquids is a demanding task and a simple and
efficient method of drying would have many applications in industry. For
example compressed air often needs to be dried before use and natural gas
is dried before entering the pipelines to avoid corrosion and pipeline
freezing problems. Compressed air dehydration is conventionally achieved
by refrigeration while natural gas is dried by packed column techniques
using either solid (molecular sieves) or liquid (glycol, amines, etc.)
absorbers [1]. More efficient drying techniques are required to reduce gas
drying costs.

Another application of interest is removal of water from organic
liquids. Examples include drying of alcohols (ethanol, isopropanol),
halogenated organic solvents (Freon, CHLOROTHENE*, PRELETE*, methylene
chloride), hydrocarbons (petroleum, kerosene, jet fuels) and other solvents
(benzene, toluene etc.). Dehydration of ethanol/water mixture beyond the
azeotropic point is very energy intensive via conventional distillation
methods [2]. Alternative technologies are required to economically
dehydrate ethanol. Dehydration of water immiscible solvents such as
hydrocarbons and halogenated hydrocarbons is challenging due to low
concentrations of water (50-2000 ppm). Conventional technologies such as
heat regenerated molecular sieves, activated alumina, silica gel and
calcium chloride are effective in reducing water levels in these solvents
down to a few ppm levels. However these techniques are energy intensive
and require a high degree of operator attention.

Dehydration of gases and liquids with membranes can be achieved if a
highly water selective membrane is available. Very high water selectivities are required because the available driving force for water transport
is limited by the water vapor pressure at the operating temperature. Hence
drying high pressure gases or organic liquids containing minute quantities
of water requires the use of membranes with water permeabilities several
orders of magnitude higher than that of the gas or liquid being dehydrated.
The membrane must also be insoluble in both water and the organic solvent.
The preferred membrane process for liquid dehydrations is pervaporation
[3]. The process involves direct contact of a liquid with one side of a
selective membrane. A partial pressure differential across the membrane is
maintained either by application of vacuum or sweep gas. As a consequence
of phase change, the process is limited to relatively volatile liquids and
is applicable when the permeating species is the minor component of the
liquid mixture. The pervaporation process has been extensively studied for
dehydration of ethanol [4-6]. Advantages of pervaporation over the conventional distillation are obvious when the vapor phase composition vs. liquid
phase composition graph (Figure 1) of a selective membrane is compared with
the vapor-liquid equilibrium (distillation) curve. This technology has
been developed to a stage where at least one membrane dehydration process
is now commercially available [7].

---

*Trademark of The Dow Chemical Company.

Ionic polymers with carboxylic and sulfonic acid functional groups have extremely high water permeabilities. Carboxy methyl cellulose/poly-acrylic acid blends show good water permeability and excellent selectivity for water over low alcohols (Table 1). These membranes can be effectively used to dehydrate alcohols [8]. Selectivity for water over alcohol increases with decreasing polarity of the alcohol. The water fluxes can be correlated with water vapor pressure and are not a function of the nature of the alcohol.

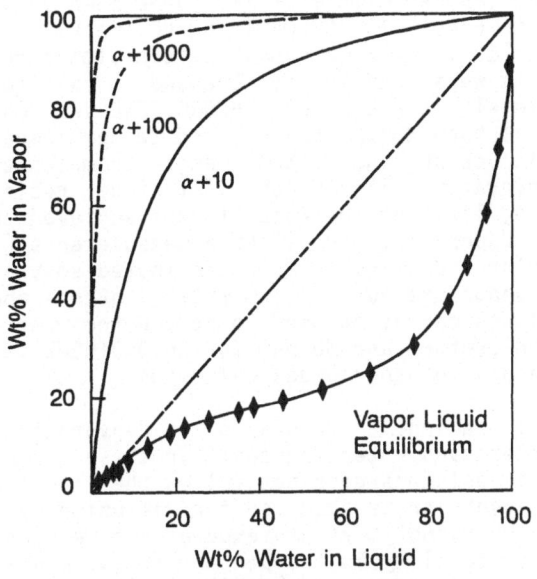

Figure 1

Vapor - liquid composition curves for ethanol-water mixture; α is the membrane selectivity for water over ethanol (8).

Table 1

WATER SELECTIVITY FROM AQUEOUS/ORGANIC SOLUTIONS

11 wt% Water/89 wt% Organic CMC/PAA (Na) 25°C (Ref.8)

Organic	Pw mmHg	Total Rate g-mil/hr-m²	Permeate Composition % Yw	Water Selectivity αw
Ethanol	11.0	50	99.7	2,700
2-propanol	15.6	155	99.99	>10,000
t-butanol	17.1	224	99.99	>10,000
1-propanol	17.3	250	99.99	>10,000
1-butanol	20.3	412	98.9	730
Acetone	16.1	424	99.853	5,500

Perfluorinated ionomers functionalized with sulfonic acid groups (i.e., Nafion and Dow's PFSA) have been extensively studied and are used in a variety of applications [9]. The hydrophobic fluorocarbon backbone imparts good chemical stability, while sulfonic acid groups make it hydrophilic. The sulfonic acid groups have been postulated to form interconnected clusters, giving these polymers unique morphology [10,11]. The ionic cluster dimensions are a function of water content. Water transport presumably occurs through swollen ionic clusters, and hence, the rate of water transport is a function of both charge density and water content of the membrane. Ionic charge density can be varied during the copolymerization process.

Perfluorinated ionomer membranes [12] have been used to dehydrate the chlorinated solvents Freon 113, CHLOROTHENE VG, methylchloroform, and methylene chloride [13]. Extremely high selectivities for water over the chlorinated solvents were observed for the sodium salt form of the membrane. Water levels in Freon 113, CHLOROTHENE VG, and methylchloroform were reduced to less than 2 ppm, 80 ppm, and 20 ppm respectively. Some loss of polar additives such as methanol and nitromethane can occur during the process of dehydration. Significant diffusional resistance was observed in the liquid phase, limiting the water fluxes achieved through the membrane [13,14]. Vapor-liquid equilibria measurements [14] showed that the presence of polar additives in the chlorinated solvents significantly reduced the water vapor pressure at low water contents and hence membrane dehydration techniques cannot be used to reduce the water levels below 20 ppm for methylchloroform and 80 ppm for CHLOROTHENE VG. Similar restrictions may apply to other organic liquids.

To effectively use ionomer membranes for dehydration applications it is necessary to understand water transport in these polymers. Molecular diffusion in swollen polymers does not follow the classical Fickian behavior. Fickian behavior is observed for diffusion of gases at low pressure through rubbery polymers at temperatures well above T_g. Under these conditions permeability is independent of gas pressure. Glassy polymers show pressure dependent permeabilities. These effects disappear at higher pressures and can be explained by dual mode theory. Similarly, permeabilities of vapors such as water in hydrophobic or mildly hydrophilic membranes are independent of water vapor pressure.

Non-ideal transport becomes evident if the permeating molecule plasticizes the membrane. Plasticization is normally induced by the high solubility of the permeating molecule in the membrane which results in swelling of the membrane. Both diffusion coefficients and solubility coefficients may be concentration dependent. Examples of such plasticization include CO_2 in cellulosics [15], organics such as toluene in silicones [16], and water in hydrophilic polymers.

Membrane-penetrant systems which show pressure (concentration) dependent permeation rates have been shown to develop highly non-linear concentration and activity profiles within the membranes [17]. Time is required to establish these profiles. In addition, water and other condensible vapors create large swelling gradients from the fully swollen wet (feed) side to unswollen dry (permeate) side of the membranes. The relaxation of the resulting stresses may control the time to attain steady-state flux [18].

Slow relaxation processes appear to be the dominant factors causing the long times required to reach steady state permeation rates. Transient permeation experiments would yield incorrect diffusion coefficients for membrane materials exhibiting this behavior. Relaxation processes, highly concentration dependent diffusion coefficients and solubility coefficients, therefore, require a more detailed approach to studying transport. This paper describes a preferred method of analyzing diffusion process.

EXPERIMENTAL

Membrane Preparation

PFSA membranes were fabricated in house according to the previously described procedures [12]. The acid form of PFSA was neutralized with NaOH to obtain the sodium salt form. Different counter ion forms were obtained by soaking the Na salt in 1.0M salt solutions of appropriate cations. The membranes were repeatedly washed with water before use. Nafion films were purchased from Aldrich.

Permeability Measurement

Water permeabilities were measured using a mass spectrometric based system (Sketch 1). Dry nitrogen gas was mixed with humidified nitrogen to get the desired water vapor pressure. A gas flow of 500 cc/minute was passed over the face of the membrane. Vacuum was used to create the driving force for the permeation. All the measurements were carried out at 30°C unless otherwise noted. Membrane thickness was measured under ambient conditions on the actual membranes used in the experiments. Thickness was measured at five different points and the average used for all calculations. The actual membrane thickness during permeability measurements may have varied slightly from these averages. Variations of the membrane thickness are not expected to be more than ±5% of the averages which is within the experimental error of ±10% estimated for the permeability data.

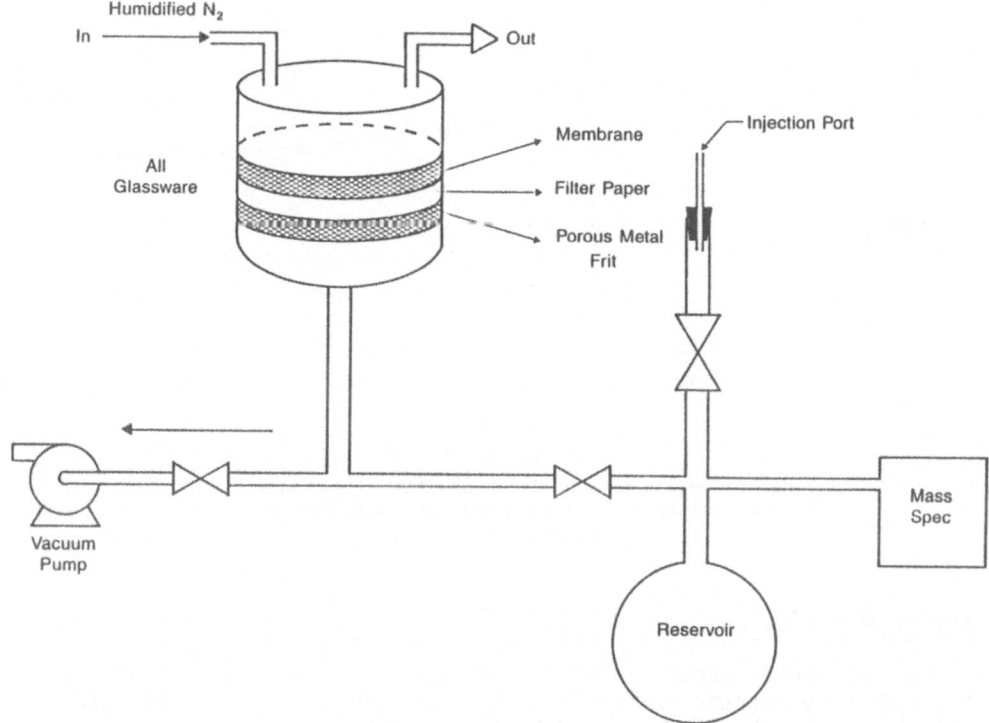

Sketch 1 Schematic of permeability measurement apparatus.

To study the down stream vapor pressure effects that exist when permeating into non-zero water vapor pressure, nitrogen sweep gas of desired water vapor pressure was used on the permeate side. The humidity of the sweep and feed gas exiting the membrane device was measured to estimate the water permeability.

Sorption Measurements

Membranes were dried under vacuum at 120°C and placed into closed bottles containing various water vapor pressures. Different saturated salt solutions were used to obtain the desired water vapor pressure. Solubility coefficients were obtained from the gravimetric measurement of the water sorbed into the membrane.

Transient Effects in Permeation Measurements

The transient effects involved in measurement of the permeabilities in ionomer membranes are illustrated in Figure 2. Membranes stored under ambient condition (about 50% relative humidity) were exposed to humidified gas upstream while vacuum was applied on the permeate side. Permeation rates were measured as a function of time. Typical data are shown in Figure 2 for films of 100 micron thickness. About 48 hours were required to achieve steady state for these membranes. Longer times were required for thicker films.

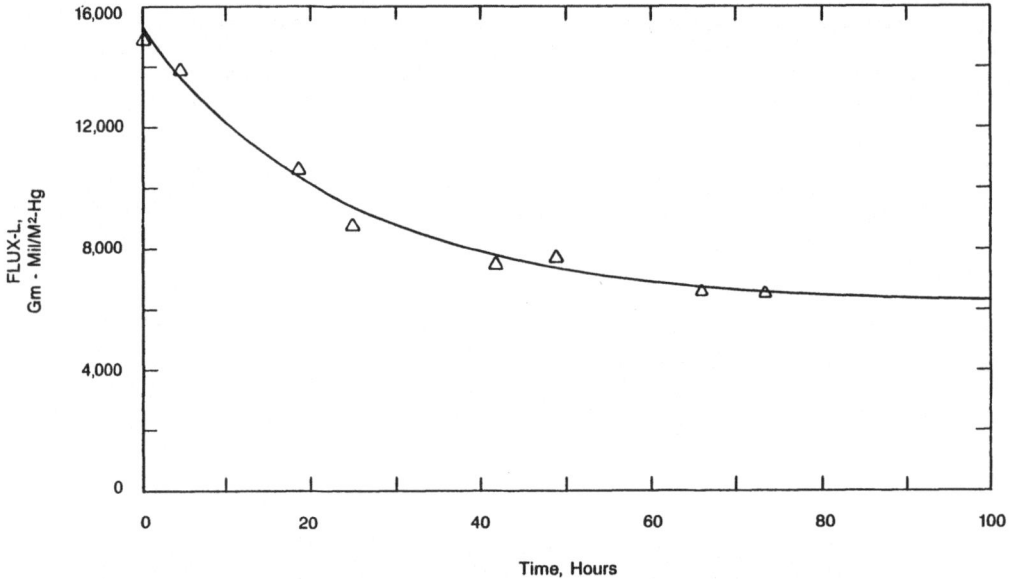

Figure 2
Transient effects in permeability measurements. Liquid water flux of PFSA membrane as a function of time. Membrane thickness 4 mils and temperature 29°C.

Analysis of Diffusion Coefficients

For ionomeric systems in which the strong interactions between ionic sites and the penetrant (water) result in concentration dependent diffusion coefficients, the Fickian diffusion model, which assumes the solubility coefficient is independent of the concentration, is not valid. The commonly used Fickian diffusion constants actually contain mobility and solubility gradient contributions [19]. In addition, the concentration dependent solubilities lead to nonlinear concentration profiles during steady state diffusion. Therefore, mobility measurements which generate average diffusion coefficients are generally not satisfactory.

The approach taken in this work attempted to eliminate the effects of relaxation, non-constant solubility coefficients, and non-linear concentration profiles. Steady state permeation experiments were performed using a broad range of feed vapor pressures. Solubility data were obtained by monitoring weight gain when samples were exposed to varying partial pressures of penetrants. Combining these two sets of data allowed a determination of the "thermodynamic" diffusion coefficient (not Fickian) as described below. The overall approach combines the concept of a thermodynamic diffusion coefficient with the variable feed partial pressure method of Rouse [20].

Flux, in general, is proportional to a gradient in chemical potential as described [3] by Equation (1).

$$J = -(c/r) \text{ grad } \mu \tag{1}$$

where c is the local concentration, r is a thermodynamic frictional resistance coefficient, and μ is the chemical potential. Chemical potential can be written as:

$$\mu = \mu_o + kT \ln P \tag{2}$$

Substituting Equation (2) into (1), and recognizing kT/r as the thermodynamic diffusion coefficient and c/P as the solubility coefficient S, gives:

$$J = -D_T(P)S(P) \text{ grad } P \tag{3}$$

$D_T(P)$ and $S(P)$ are the thermodynamic diffusivity and solubility coefficients, respectively. Both are dependent on the local concentration of the penetrant or its corresponding vapor pressure, P. Integrating across the membrane under steady state conditions gives:

$$JL = \int_{P_{lo}}^{P_{hi}} D_T(P) \; S(P) \; dP \tag{4}$$

P_{hi} and P_{lo} are the feed and permeate penetrant partial pressures respectively. L is the membrane thickness. $D_T(P)$ and $S(P)$ are still the vapor pressure dependent thermodynamic diffusivity and solubility. Differentiating Equation (4) with respect to P_{hi} gives:

$$\frac{\partial(JL)}{\partial P_{hi}} = D_T(P_{hi})S(P_{hi}) \tag{5}$$

Data obtained under steady state conditions where P_{lo} is kept constant, i.e., equal to zero (vacuum), and P_{hi} is varied can be analyzed using Equation (5). Since $S(P)$ (equal to $c(P)/P$) can be obtained from equilibrium solubility measurements, it is possible to determine $D_T(P)$. $D_T(c)$ can also be obtained since the solubility data gives the relationship between c and P.

The method described above allows the thermodynamic diffusivity D_T to be determined. Unlike the Fickian diffusivity, the thermodynamic one is related directly to mobility and has no anomalous corrections which must be applied because of a solubility coefficient gradient across the membrane.

In addition, this diffusivity describes the mobility within a differential thickness element inside a membrane rather than an average across the entire membrane. This diffusivity is specified for a specific penetrant concentration or partial pressure. It corresponds to what would be obtained from an experiment using an infinitesimal partial pressure drop across the membrane.

RESULTS AND DISCUSSION

Water permeability data measured for a number of polymers as a function of water vapor pressure are given in Figure 3. Water permeability is essentially independent of water vapor pressure for hydrophobic polymers such as silicone, polysulfone etc. Water swellable polymers (e.g., ionomers such as Nafion and PFSA show increasing water permeability with increasing water vapor pressure. Cellulosic polymers also show a similar effect, albeit to a lesser degree.

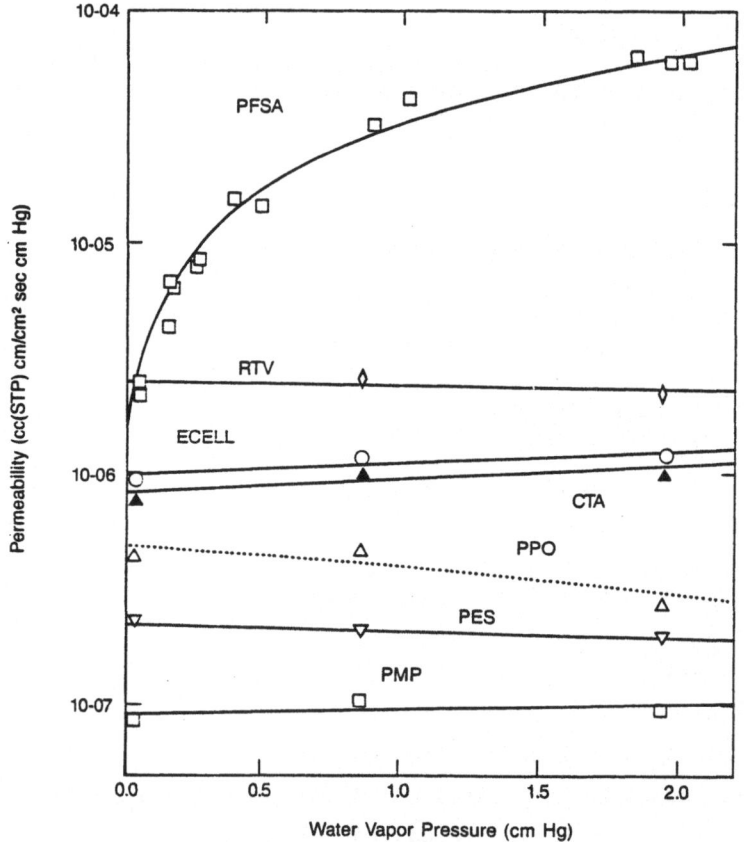

Figure 3
Water permeabilities of various polymers at different water vapor pressures.
Temperature 30°C.

The extremely high water permeabilities of the perfluorinated ionomers at high vapor pressures illustrate vividly why these polymers are of interest as dehydration membranes. At 2 cm Hg, water permeates through these materials nearly 100 times faster than in cellulosics. At very low vapor pressure, however, performance is similar to other polymers.

To obtain the thermodynamic diffusion coefficients, equilibrium sorption of water was measured in PFSA at different vapor pressures (Figure 4). At low vapor pressures there is a significant uptake of water due to initial hydration of ionic sites. As vapor pressure increases, the water uptake continues to increase, although at a slower rate. At high relative humidities there is an upturn in the sorption isotherm which corresponds to significant swelling of the polymer. The solubility coefficient, c(P)/P, obviously is not constant. Initially it is very high due to the strong ionic site interactions. At higher relative humidities it decreases significantly. This illustrates the problem with using a Fickian diffusion analysis. A constant solubility coefficient is not even a reasonable rough approximation.

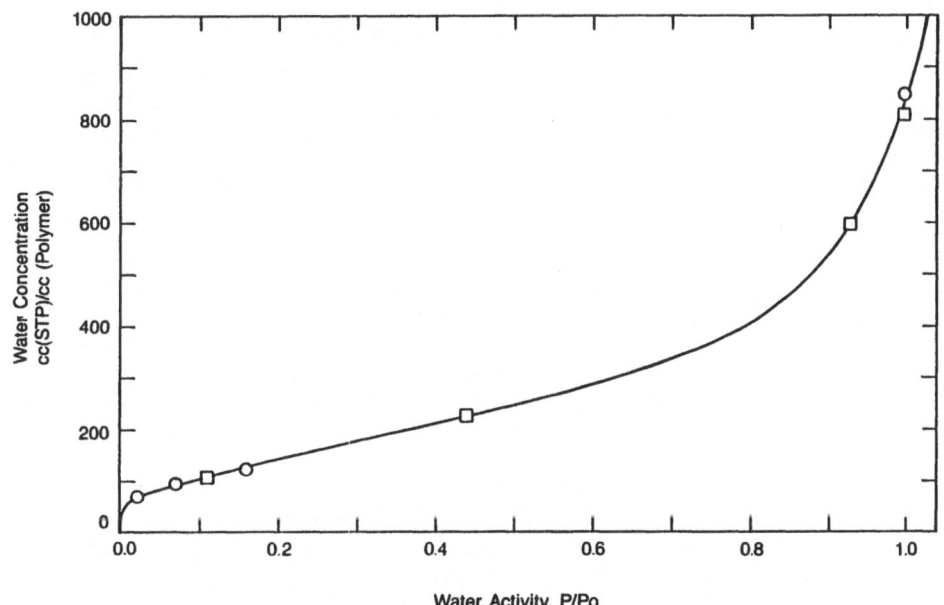

Figure 4
Water vapor sorption into PFSA at different water activities.
25°C and   55°C.

Steady state permeation data for water in PFSA are shown in Figure 5. As the water vapor pressure of the feed increases, the flux increases by over three orders of magnitude. Differentiating the curve of Figure 5 as described by Equation (5) generates a differential permeability. From the solubility data given in Figure 4 the thermodynamic diffusivity was calculated. This result is shown in Figure 6.

Initially the solubility coefficient is very high due to strong ionic site interactions with water. These strong interactions, however, hinder the mobility of water, resulting in low diffusivity. After the initial hydration of the ionic sites occurs, the additional water absorbed is much more mobile. This produces the strong upturn in diffusivity shown in Figure 6. At very high water vapor pressures, the polymer is highly swollen, and the mobilities approach the self diffusion of water in water. Tortuosity of the ionic domain channels probably causes the diffusivity to approach a limiting value which is somewhat lower than that of self diffusion of water in water.

Once $D_T(P)S(P)$ is determined from Equation (5), Equation (3) can be integrated to give the effective partial pressure P as a function of position in the membrane. The solubility data can be used to convert the effective partial pressure of water at each point in the membrane to a local concentration. Figure 7 shows the water concentration profiles obtained in this way for a PFSA membrane at 60°C with vacuum on the permeate side. Four different water activities are shown. (Activity 1 being liquid water.) For all activities, a very thin "dry skin" exists on the permeate side. Most of the membrane has a very gradual concentration gradient. At high feed water activities, however, the membranes swells considerably near the feed side. This produces a very sharp increase in concentration near the feed under those conditions.

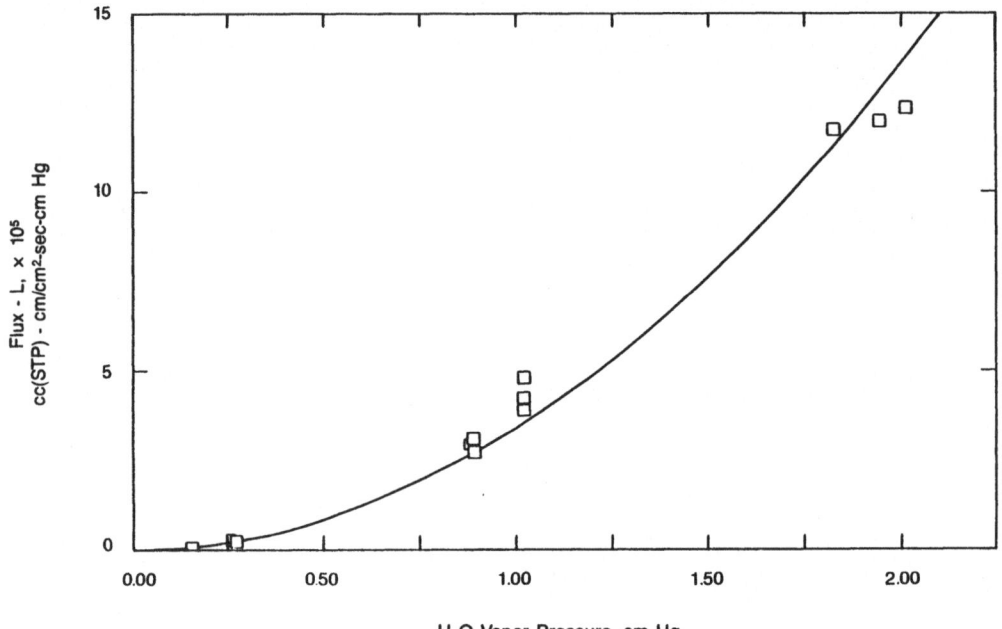

Figure 5
Water flux of PFSA membrane at different water vapor pressures.
Membrane thickness 4 mils.

Under real world use conditions, a membrane rarely has near-zero water vapor pressure on its permeate side. Equation (4) allows the flux (J) to be calculated from D(P)S(P) obtained via Equation (5). Experiments were performed to verify these predictions. A flowing feed and sweep experiment was carried out using variable relative humidities for both. The experimental data are shown in Figure 8 along with the predicted performance shown by the solid lines. Agreement is very good. The curvature is characteristic of materials which have concentration dependent diffusion coefficients. Although many researchers use an exponential function for D(C), it is clear from Figure 6 that such a function is not appropriate for these membranes. We have found that the product $D_T(P)S(P)$ is modeled well by setting it proportional to $P^{1.8}$.

THICKNESS EFFECTS

It is generally assumed that flux is inversely proportional to membrane thickness. Very early on in this study, it became obvious that ionomers deviate from this behavior. Figure 9 shows data for permeability (normalized to thickness) as a function of thickness. It is evident that the thicker films show higher fluxes per unit thickness.

Thickness effects on permeability can be analyzed by Equation (6)

$$1/\underline{P} = 1/\underline{P}_0 + (r_1 + r_2)/L \qquad\qquad (6)$$

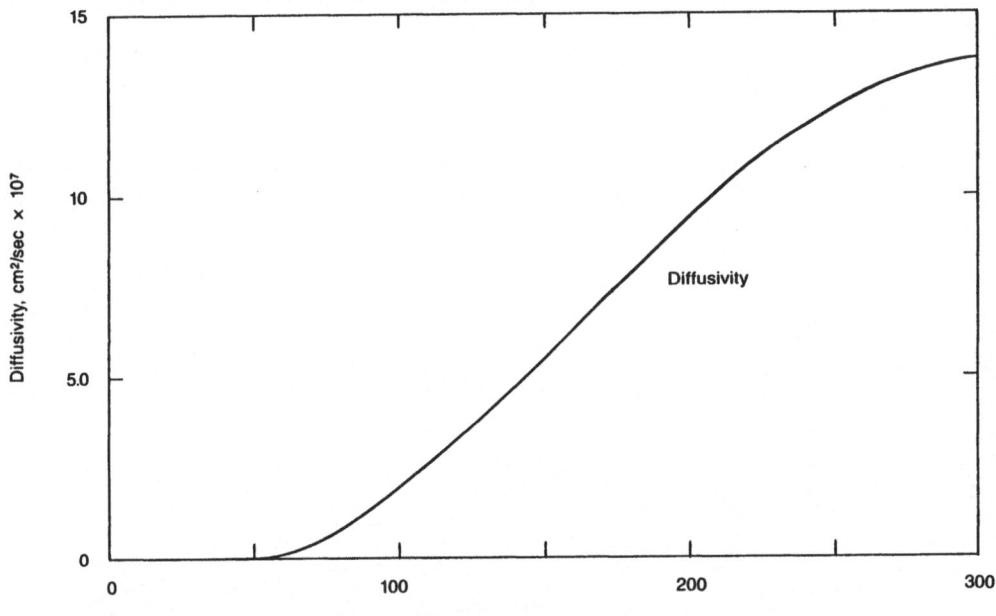

Figure 6
Diffusion coefficient of water at different water contents of PFSA.
Temperature 30°C.

where $\underline{P}$ is the mean permeability coefficient averaged over a membrane of thickness L, $\underline{P}_0$ is the mean permeability coefficient at infinite thickness, and $(r_1 + r_2)$ is the sum of resistances at the membrane interfaces [21]. Figure 10 shows a plot according to Equation (6), of permeability data from Nafion films of various thickness. The slope of these curves, which is equal to the sum of interfacial resistances, increases with decreasing water vapor pressure. The permeability at infinite thickness, obtained from the intercepts, increases with water vapor pressure.

The cause of this interfacial resistance in these membranes remains unclear, but several mechanisms have been considered. These include:

(1) External boundary layers or resistances at either interfaces

(2) Skin effects

(3) Temperature effects

(4) Interfacial resistance which prevents attainment of the equilibrium sorption concentration in the membrane at steady state.

There is evidence against all except number (4).

An external boundary layer resistance due to poor mixing at the feed-membrane interface, especially for high selectivity, high flux membranes, could result in a boundary layer resistance from water vapor depletion. However, varying the feed rates did not change the water flux as would be expected if this mechanism were operative. Further, as seen in Figure 10, the thickness effect is greater at low vapor pressures where flux is lower, a condition less likely to cause this kind of boundary layer. Boundary layers of this nature are common in liquid phase systems [22], where diffusion coefficients are about five orders of magnitude lower than in the gas phase.

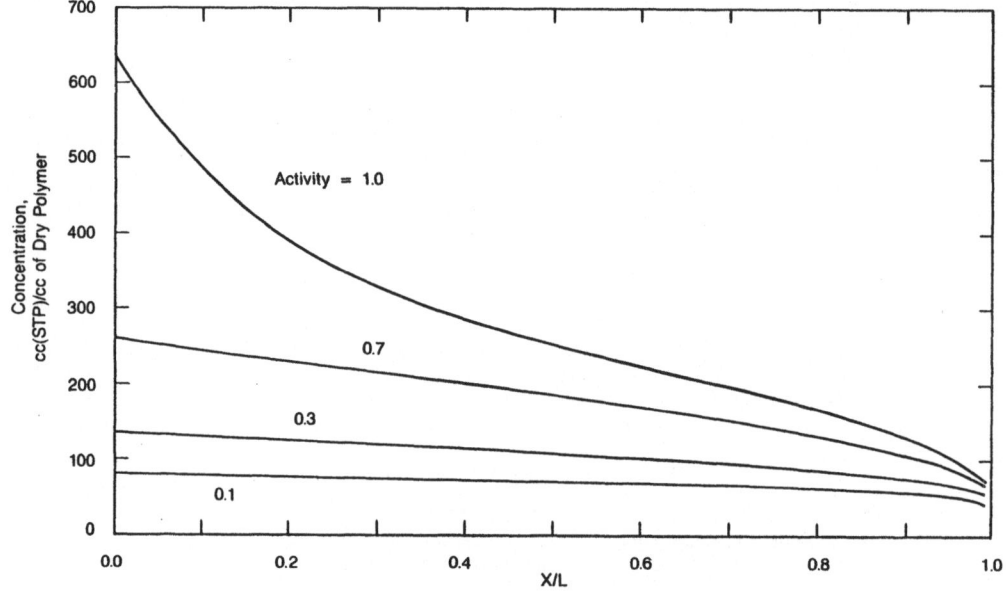

Figure 7
Water concentration profiles in PFSA at 60°C.
X/L is the fraction of the membrane thickness.

The question of resistances on the low pressure side of the membrane has also been addressed. Pressure drops across filter paper and porous metal frit membrane supports used in these experiments were measured and found to be not large enough to account for the thickness effects [23]. Again, the greatest effect of this type of resistance is at high fluxes, and the largest thickness effect is at low flux conditions.

Thickness effects on permeability have been attributed to a high resistance surface "skin" [24]. Such systems can be treated as a laminate of two different films where permeability changes with the relative thickness of two layers. Such effects can be ruled out in the present case, because the effect was observed when the thickness was varied by stacking the membranes and the ratio of skin thickness to core thickness should have remained constant.

It is interesting to note that thickness effects have also been observed for a series of organic vapors through polyethylene films [25]. This behavior is therefore not unique to water transport or to ionic polymers, and a general mechanism may be applicable.

Temperature effects of permeation and sorption have been reported to be significant [26]. The possibility that a temperature differential across the membrane under steady state operation could be responsible for the thickness effect was considered. The heat of solution of water is negative (exothermic) so that temperature might increase at the feed interface and decrease at the low pressure interface. Calculations using the highest water flux and the thermal conductivity of Teflon have shown that only an extremely small temperature differential should be present and is not likely to be the cause of the thickness effect. The concentration profiles of permeating vapors in films during steady state permeation have been experimentally determined [27]. The concentration of the permeating species in the membrane at the feed/membrane interface is often, but not always, found to be less than the static equilibrium concentration. This shows that a resistance can exist at the feed/membrane interface which would yield a thickness effect of the nature observed in the present work. The cause of this resistance remains a matter of speculation.

However, the increase in resistance with decreasing water vapor pressure in the present case appears to be a new clue and has been useful in at least arguing against some possible mechanisms. It appears feasible that the surface properties of the film are changing in response to the external environment, specifically the relative humidity. There is evidence that a surface of a methacrylate hydrogel is more hydrophobic in air saturated with water vapor than in liquid water [28]. If we assume that ionomers show a similar behavior, and that the surface becomes still more hydrophobic at lower humidities, then greater resistance to water sorption could result, thus explaining the observed results.

There are serious practical consequences for the observed thickness effect. It is common practice to reduce the thickness of the discriminating layer of a membrane to maximize the productivity of a membrane device. This is achieved either by using a thin film coating on a porous substrate or using an asymmetric membrane structure. As a consequence of the thickness effect these approaches may not be applicable to the dehydration membranes made from highly water permeable ionomers. Similar restrictions will apply to other membrane separations exhibiting similar thickness effects.

CONCLUSIONS

Perluorosulfonated ionomer membranes have high water permeabilities and excellent selectivities for water over most gases and organic liquids. These membranes have been shown to be useful for dehydration applications. Water transport in these membranes is non-Fickian. Water permeability and solubility coefficient vary with water vapor pressure. Concentration dependent thermodynamic diffusion coefficients have been obtained by combining the steady state water permeability data and the equilibrium sorption data. These diffusion coefficients correspond to what would be obtained from an experiment with infinitismal partial pressure drop across the membrane.

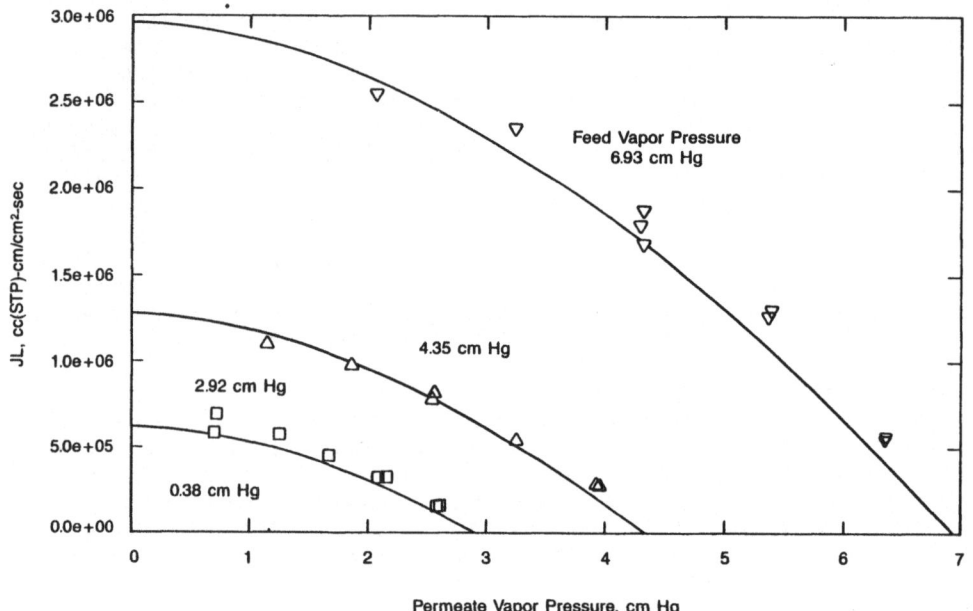

Figure 8
Effect of down stream water vapor pressure on water flux
of PFSA at different feed water vapor pressures at 60°C.

At very low water vapor pressure strong ion-water interactions induce large solubility coefficients. These interactions also result in hindered mobility and hence low diffusivity. Additional sorbed water (a = 0.1 to 0.6) is much more mobile but has a smaller solubility coefficient. At water activities $\geq$0.6 water diffusivity reaches a limiting value while the solubility coefficient increases rapidly.

Anomalous thickness effects were observed. Thickness averaged permeabilities increased with increasing film thickness. A plausible mechanism for this effect involves the interfacial resistance which prevents attainment of the equilibrium sorption concentration in the membrane under steady state conditions. More detailed characterization of membrane surfaces under different water vapor pressures will be required to better understand these interfacial effects.

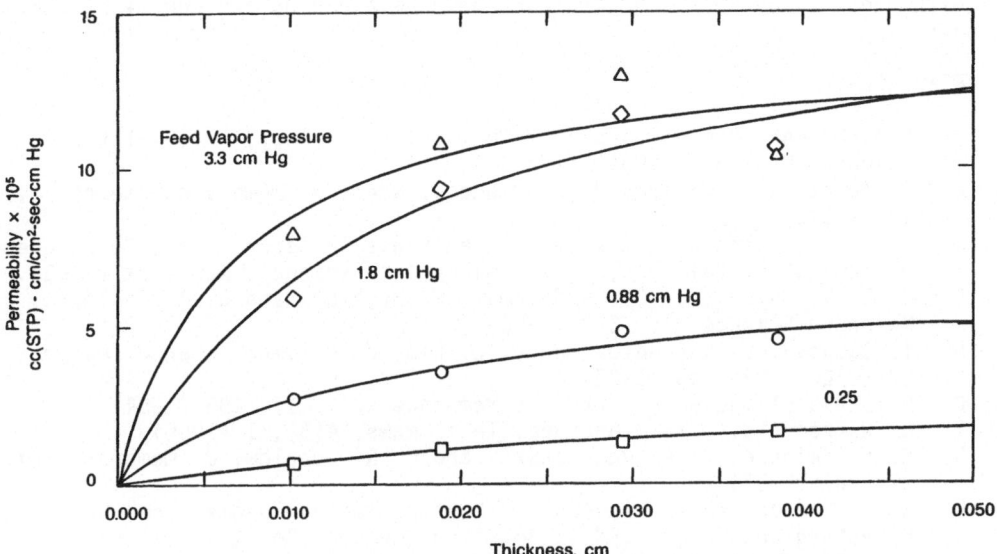

Figure 9
Thickness effects on water permeability of Nafion at different
water vapor pressures at 30°C.

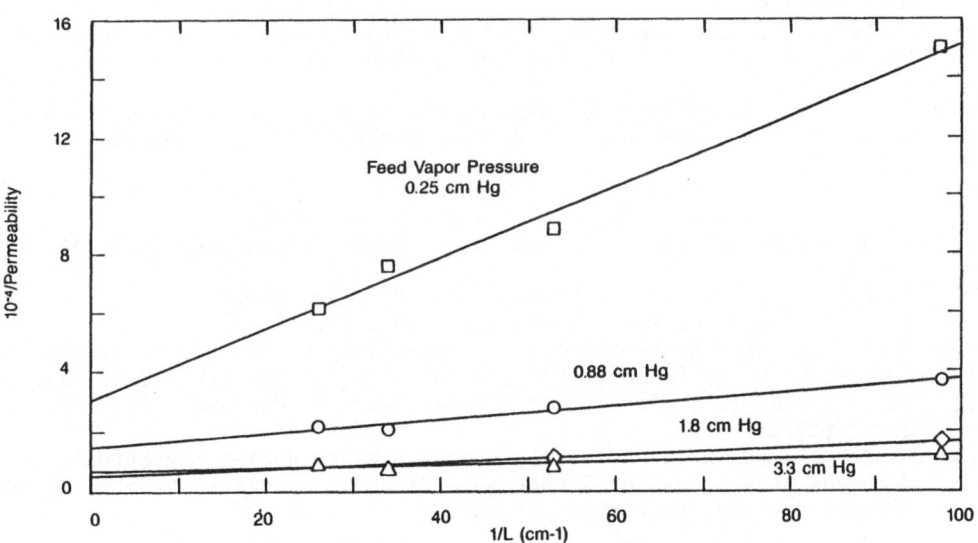

Figure 10
Plot of 1/P vs 1/L for Nafion at 30°C according to eqn (6).
Permeability units are cc(STP) - cm/cm²-sec-cm-hr)

ACKNOWLEDGEMENTS

We wish to thank Terry Gordon and Charlie Martin for providing the variable down stream vapor pressure data. We also wish to thank Keith Denslow and Rick Lundgard for their help during the course of this work.

REFERENCES

1. Kirk-Othmer, "Encyclopedia of Chemical Technology," Wiley-Interscience, New York (1984), p.p.114.
2. P. Aptel, N. Challard, J. Cuny and J. Neel, J. Membrane Science, 1, 271 (1976).
3. S. T. Hwang and K. Kammermayer, Membranes in Separations, Techniques of Chemistry, Vol. VII, p. 99, Wiley-Interscience, New York (1982).
4. M. H. V. Mulder, J. O. Hendrikman, H. Hegeman, and C. A. Smolders, J. Membrane Science, 16, 269 (1983).
5. I. Cabasso, E. Korngold, and Z. Z. Liu, J. Polymer Science, Polym. Lett. Ed., 23, 333 (1985).
6. P. Schissel and R. A. Orth, J. Membrane Sci., 17, 109 (1984).
7. D. Warren (Ed.), Membrane Sep. Tech. News, 4(5), 1 (1986).
8. C. E. Reineke, J. A. Jagdzonski, and K. R. Denslow, J. Membrane Sci., 32, 207 (1987).
9. H. L. Yeager and A. Eisenberg, "Perfluorinated Ionomer Membranes," Eisenberg and Yeager (Ed.), American Chemical Society Symposium Series No. 180, Washington, D.C., 1982.
10. T. D. Gierke, "152 Meeting of Electrochemical Society," Atlanta, GA (1977).
11. T. D. Gierke, G. E. Munn, and F. C. Wilson, J. Polym. Sci. Polym. Phys. Ed. 19, 1687 (1981).
12. B. R. Ezzell, W. P. Carl, and W. A. Mod, U.S. 4,358,545 assigned to The Dow Chemical Company.
13. D. Reddy and C. E. Reineke, "Second International Conference on Pervaporation Processes in the Chemical Industry," San Antonio, TX, March 8, 1987.
14. D. Reddy and C. E. Reineke, "New Membrane Materials and Process for Separation," Sirkar and Lloyd (Ed.), AICHE Symposium series 261 (1988), 84.
15. E. Sanders, AIChE National Meeting New Orleans, LA, April 6, 1986.
16. R. W. Baker, N. Yoshioka, J. M. Mohr and A. Khan, J. Membrane Sci. 31, 259 (1988).
17. S. N. Kim and K. Kammermeyer, Separation Sci. 5(6), 679 (1970); M. H. V. Smolder, J. Membrane Sci., 17, 284 (1984).
18. D. A. Blackadder and J. S. Keniry, J. Appl. Polymer Sci., 17, 351 (1973).
19. A. Peterlin, Colloid and Polymer Sci. 263, 35 (1985).
20. P. E. J. Rouse, J. Amer. Chem. Soc., 69, 1071 (1947).
21. S. T. Hwang and K. Kammermeyer "Permeability of Plastic Films and Coatings," H. Hopfenberg (Ed.), Plenum Press, New York (1974), 197.
22. S. T. Hwang, T. E. Tang and K. Kammermeyer, J. Macromol. Sci. Phys., B5, 1 (1971).
23. D. J. Moll and C. E. Reineke, "Porous Support Resistance Effects in Membrane Separations of Condensable Vapors," presented at the ACS New Orleans Meeting, September 1, 1987.
24. E. Sacher and J. R. Susko, Appl. Polymer Sci., 27, 3893 (1982).
25. V. L. Simril and A. Hershberger, Modern Plastics, 97 (1950).
26. R. Rautenbach and R. Albrecht, J. Membrane Sci., 7, 203 (1980).
27. M. H. V. Mulder, A. C. M. Franken, and C. A. Smolders, J. Membrane Sci., 23, 41 (1985).
28. F. J. Holly and M. F. Refojo, J. Biomed. Mater. Res., 9, 315 (1975).

THE STRUCTURE OF ASYMMETRIC HOLLOW FIBER MEMBRANES

BY OXYGEN PLASMA ABLATION

A. K. Fritzsche

PERMEA, INC., A Monsanto Company
11444 Lackland Rd.
St. Louis, MO 63146

INTRODUCTION

Loeb and Sourirajan invented the first integrally-skinned membrane in 1960 for desalination by phase inversion of cellulose acetate sols (1). In the interrally-skinned membrane, the skin and substructure are composed of the same material. The skin layer determines both the permeability and selectivity of the bilayer, whereas the porous substructure functions primarily as a physical support for the skin. Differences in density between the two layers are the result of interfacial forces and the fact that solvent loss occurs more rapidly from the air-solution and solution-nonsolvent bath interfaces than from the solution interior (2).

Integrally-skinned hollow fiber membranes suitable for commercial gas separations were subsequently developed by applications of a special coating procedure (3). A thin layer of a highly permeable, nonselective polymer was applied to the surface of an asymmetric polysulfone hollow fiber membrane. This coating sufficiently reduces the permeability through the pores and defects within the skin to render permeation through the normal skin layer predominant. As in the cellulose membrane, the internal matrix of the hollow fiber does not affect gas separation, but it does offer resistance to passage of gases through the hollow fiber.

Exposure of these integrally-skinned hollow fiber membranes to an oxygen plasma has proven itself a viable procedure for investigating the hollow fiber membrane structure. With electrodeless radiofrequency excitation, diatomic oxygen molecules are disassociated into free radicals and ions. At low pressures, these moieties are sufficiently long-lived to react with a variety of polymers, which is reflected by a loss of weight with time (4,5). The rate of ablation is dependent upon the chemical structure of the polymer as well as the radiofrequency power level and the oxygen flow rate (4-6). The plasma treatment has been found to be confined to the surface regions of a polymer and does not alter its bulk properties (7-11). For this reason, plasma treatment has been used to alter polymer adhesive bondability and wettability (6, 11-17) as well as to study surface morphology (8, 18-22). An additional advantage is that oxygen plasma ablation is a low temperature process (4). Therefore, the internal structure of the sample remains unaltered by thermal effects.

Oxygen plasma ablation combined with scanning electron microscopy has been used to define the structure of hollow fiber membranes prepared from polysulfone in N-formylpiperidine/formamide solvent/nonsolvent mixtures (23). These asymmetric hollow fiber membranes possess a microscopically observable skin supported by a porous open cellular matrix. The oxygen plasma ablation studies indicated that the effective separating layer of the asymmetric hollow fiber membrane is only a small fraction of the thickness of this microscopically observable skin. Below the effective separating layer, this skin contains pores and channels with dimensions below the limits of resolution of the scanning electron microscope.

Asymmetric hollow fiber membranes are formed by dissolving the polymer in a solvent and subsequently by spinning into a coagulation medium, such as water. Solvents such as dimethylformamide, dimethylacetamide, N-methylpyrrolidone and formylpiperidine can be utilized to fabricate asymmetric hollow fiber membranes from polysulfone (2,3). The selection of the solvent impacts the structure of the resulting asymmetric membrane due to their differing polymer-solvent interactions. These interactions affect the degree of orientation of the polymer chains in the dope during spinning and coagulation as well as the kinetics of coagulation. Often small amounts of nonsolvent, such as formamide, are added to the spinning dope to affect the kinetics of coagulation and concomitantly the structure of the resultant asymmetric hollow fiber membranes. Therefore, a gamut of asymmetric hollow fiber membranes may be obtained through alteration of the solvent, nonsolvent, and their relative concentrations as well as the processing conditions. Characterization of the resultant membrane structures is derived primarily from their performance and examination of their electron scanning micrographs. However, electron microscopy is limited by the degree of resolution of the instrument. The dimensions of the pores and channels in an asymmetric membrane diminish as the distance from the outer surface decreases. Examination of the scanning electron micrographs suggest a dense outer skin, which is thicker than the effective separating layer of the asymmetric membrane. Therefore, oxygen plasma ablation by successive, low temperature removal of the outer surface of the asymmetric yields information which facilitates the interpretation of scanning electron microscopical results. In this paper, the utility of this combination is demonstrated with structural studies on asymmetric hollow fiber membranes prepared by phase inversion from polysulfone, polyethersulfone, and polyphenylsulfone.

EXPERIMENTAL

Test cells were prepared from hollow fiber membranes spun by C.A. Cruse from 37% total solids polysulfone (Amoco Udel 35), 40% total solids polyethersulfone (ICI Victres 600P), and 37% total solids polyphenylsulfone (Amoco Radel A400) dopes in solutions containing 85% by weight N-methylpyrrolidone and 15% by weight formamide. Each test cell contained five hollow fibers approximately 15cm in length. One end of the collection was encapsulated in a 3/8" diameter epoxy seal. The seal fixes the cell in the gas test assembly and prohibits intermixture of the feed and product gases. The opposite ends of the hollow fiber membranes were sealed with a rapidly curing epoxy. Therefore, the only gas which can be measured emerging from the cell is that which is transported across the hollow fiber membranes. Sufficient test cells were made of each type to insure duplicate or triplicate samples of each oxygen exposure time. For example, sixteen test cells of the polyphenylsulfone hollow fiber membrane were made. These sixteen were divided into six sets. Triplicate samples were exposed to oxygen plasma for one, three, and eight minutes. Duplicate samples were similarly treated at two and five minutes. Three samples were also maintained as unexposed controls. The oxygen plasma ablation experiments were performed using a Plasma Technology System 80 Barrel

Etcher. A radiofrequency of 13.6 megahertz was maintained at 50 watts during these experiments. The oxygen flow rate was 20 sccm (standard cm³/min), and the temperature within the etcher remained in the 27-28°C range throughout all of these experiments.

The oxygen plasma ablated samples and the unetched controls were measured for their uncoated helium and nitrogen flux rates at 100 psig at room temperature. The helium and nitrogen permeabilities were calculated by equation 1.

$$P/\ell = \frac{Q}{A\Delta P} = \frac{Q}{n\pi DL\Delta P} \tag{1}$$

where

P = the permeability of the separating layer $cm^3(STD)$-cm/ cm^2-sec-cmHg,
ℓ = the effective thickness of the separating layer,
Q = the gas flux,
A = the surface area of the membrane,
n = the number of fibers in the sampling,
D = the outer diameter of the hollow fiber membrane,
L = active length of the fibers, and
ΔP = differential pressure between outer surface and bore of the hollow fiber membrane.

The separation factor, also referred to as the selectivity of the gas pair, is the ratio of their permeation rates. For a gas pair, such as helium and nitrogen, the separation factor is given by equation 2.

$$\alpha = \frac{P/\ell(He)}{P/\ell(N_2)} \tag{2}$$

Samples were also examined by C.J. Pellegrin with a JEOL JSM-840 scanning electron microscope. This instrument has a resolution of 4nm, and photomicrographs with magnifications up to 50K were obtained on the submitted samples.

RESULTS AND DISCUSSION

Polysulfone: A scanning electron micrograph of a cross-section of an asymmetric polysulfone hollow fiber membrane spun from a solution of N-methylpyrrolidone (84 wt%)/formamide (16 wt%) is shown in Figure 1. This asymmetric hollow fiber membrane consists of a dense outer skin, which is not discernible at this magnification, subtended by a porous matrix. This internal matrix, shown in Figure 2, is an open-cellular structure in which the cell walls are composed of agglomerates of spherical micelles. This open-cellular, three dimensional micellar matrix provides little resistance to gas flow but instead serves to support the dense, outer layer which contains the effective separating layer and to maintain the structural integrity of the asymmetric hollow fiber membrane at elevated pressures. As the outer surface of the asymmetric hollow fiber membrane is approached, the spherical micelles become more tightly packed and the interstitial cells smaller, as shown in Figure 3. In the outer skin, the micelles become so tightly packed that it becomes difficult to identify their boundaries. A cursory examination of this figure indicates that the dense skin of this hollow fiber membrane is approximately 1.2-1.5µm (1200-15nm) thick.

This dense skin is seen in greater detail in Figure 4. The tightly packed array of spherical micelles is observed whose packing density increases as the outer surface layer is approached. At the outer surface, the micellar structure is indistinct. However, channels can be observed

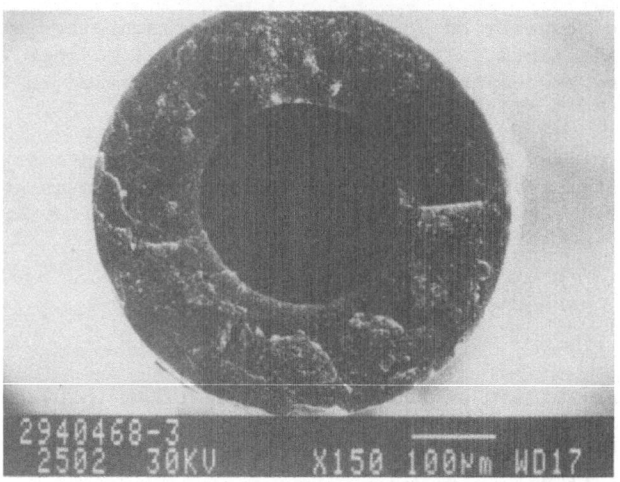

Fig.1 SEM Photomicrograph of a Cross-section of an Asymmetric Polysulfone
Hollow Fiber Membrane Spun from N-methylpyrroildone/Formamide at 150
Magnification

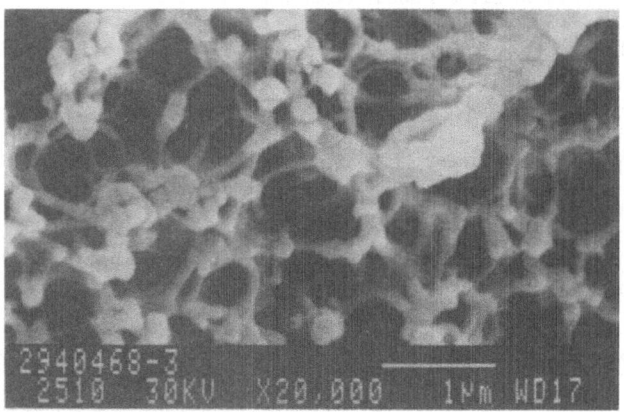

Fig.2 SEM Photomicrograph of the Open-cellular Matrix of an Asymmetric Polysulfone Hollow Fiber Membrane Spun from N-methylpyrrolidone/Formamide
at 20K Magnification

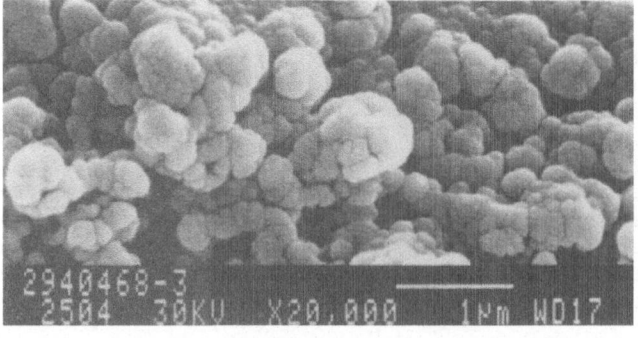

Fig. 3 SEM Photomicrograph of the Outer Edge of an Asymmetric Polysulfone
Hollow Fiber Membrane Spun from N-methylpyrrolidone/Formamide at 20K
Magnification

88

which often penetrate to the surface. Oxygen plasma ablation by successive low temperature removal of surface polymer facilities the interpretation of these scanning electron micrographs. The change in gas transport performance of these asymmetric hollow fiber membranes after oxygen plasma ablation is given in Table I and illustrated graphically in Figure 5.

Fig 4   SEM Photomicrograph of the Outer Edge of an Asymmetric Polysulfone Hollow Fiber Membrane spun from N-methyl-pyrrolidone/Formamide at 50K Magnification

TABLE I

Performance of Polysulfone Hollow
Fiber Membrane Spun from
N-methylpyrrolidone/Formamide
as Function of Oxygen Plasma Ablation Time

Time (min)	$P/\ell(He)^a$	$P/\ell(N_2)^a$	α
0	31.5	1.12	34.1
1	28.0	1.30	25.0
2	27.4	1.13	26.1
2½	30.0	2.95	10.3
3	30.6	2.04	15.2
4	31.4	2.36	11.3
8	35.9	5.08	7.2
12	43.7	10.5	4.6
20	54.8	15.7	3.5
30	77.7	26.6	2.9

a All permeability units are $cm^3(STD)/cm^2$-sec-cmHg

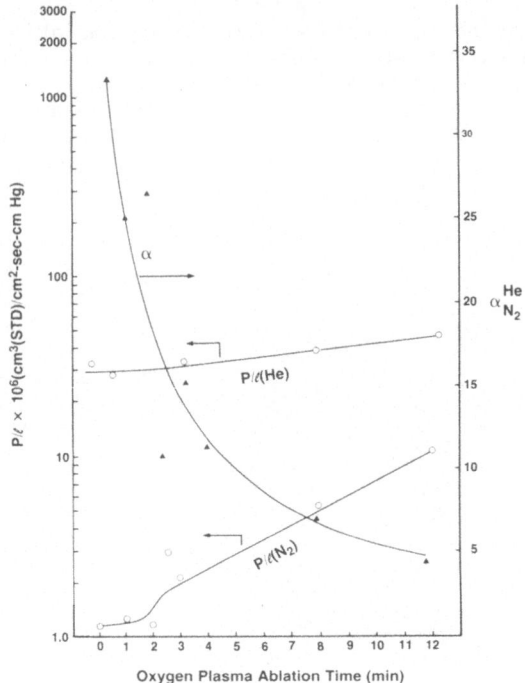

Fig 5   The Performance of Asymmetric Polysulfone Hollow Fiber
Membranes spun from N-methylpyrrolidone/Formamide as
Function of Oxygen Plasma Ablation Time

Both the the helium and nitrogen permeation rates through these hollow
fiber membranes exhibit only minor changes in the first two minutes of
etching. Subsequent oxygen plasma exposure yields gradual increases in the
helium and nitrogen permeation with the rate of increase in the nitrogen
permeability being significantly greater than the rate of increase in the
helium permeability. Using an oxygen plasma ablation rate for polysulfone
of 11.2±1.1nm/min at 50 watts RF and 20 sccm oxygen flow rate (23), these
results imply a dense effective separating layer approximately 22±2nm
thick, i.e., about 1.5-2.0% of the apparent skin thickness. The micelles
at the surface of the membrane appear, however, to be 50nm of more in
diameter from inspection of the scanning electron micrographs. Therefore,
these results also imply that these microscopically observable micelles are
themselves aggregations of yet smaller structures, which possess
microporous channels between them.

    Beneath this effective separating layer is a transitional zone
containing ever increasing quantities of micropores. The existence of
these micropores is implied by the more rapid rate of increase in nitrogen
permeability over that of helium. Helium, being a fast gas, permeates
rapidly through polysulfone in contrast to nitrogen, a slow gas. If a pore
or channel is present in the membrane, a much larger fraction of the
permeate slow gas will have traversed the pore than the fast gas. For
example, coating the membrane, which plugs surface defects and renders
membrane suitable commercial gas separations according to the teachings of
Henis and Tripodi (3), has little influence on the fast gas permeability
but results in a dramatic decline in slow gas permeability yielding
membrane with high selectivity. Therefore, if no subsurface micrpores were
exposed with oxygen plasma ablation, the flux rates of both the nitrogen
and helium would increase due to reduction in the effective separating
layer thickness, but the selectivity would remain constant. However, these
results show a decline in the selectivity with oxygen plasma exposure time,
which becomes very pronounced after two minutes of treatments.

Beyond the transitional region, the flux through the hollow fiber membrane is controlled by the structure of the supporting matrix, i.e., degree of porosity, pore size, and pore structure (open versus closed cells). However, the permeation rates and selectivities measured after thirty minutes oxygen plasma ablation indicate that the transitional zone was not traversed in these etching experiments and has a thickness greater than 350nm but less than that of the microscopically observable skin. Polyethersulfone: A scanning electron micrograph of a cross-section of an asymmetric polyethersulfone hollow fiber membrane spun from a solution of N-methylpyrrolidone (85 wt%)/formamide (15 wt%) is shown in Figure 6. In this figure, the outer skin appears to be of a thickness that it can be detected upon close inspection. In fact, an inner skin is also observed surrounding the hollow fiber bore. Between these two structures, the porous matrix, which contains two concentric rings of small macrovoids, is observed.

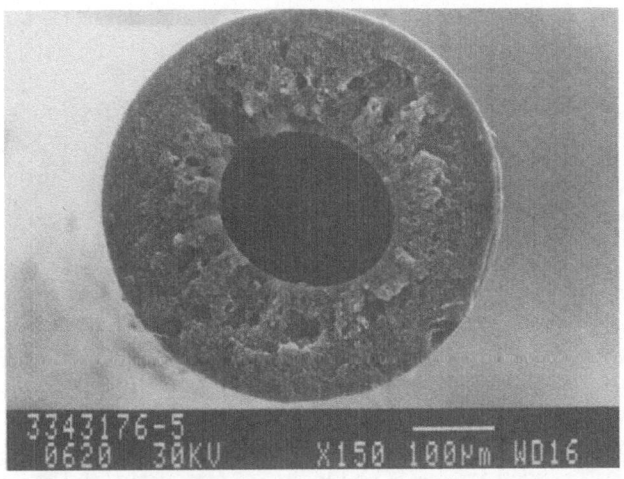

Fig. 6   SEM Photomicrograph of a Cross-section of an Asymmetric
          Polyethersulfone Hollow Fiber Membrane spun from
          N-methylpyrrolidone/Formamide at 150 Magnification

The outer edge of this asymmetric polyethersulfone hollow fiber membrane is shown in Figures 7 and 8. Examination of these figures reveals that the dense outer skin is equivalent in thickness to that of the polysulfone hollow fiber spun from the same solvent/nonsolvent mixture. However, both of the two hollow fiber membranes do appear different in both their skins and their supporting matrices. The skin of the asymmetric polyethersulfone hollow fiber membrane appears denser than that of its polysulfone congener. Also, the skin structure is more monotonous with less distinction between the impinging micelles. In the supporting matrix, the individual micelles also appear to be more tightly agglomerated suggesting a lower degree of porosity. This lower porosity, resulting from the more highly packed aggregations of micelles, gives the appearance of a thicker skin as indicated in the SEM photomicrograph of its cross-section, shown in Figure 6.

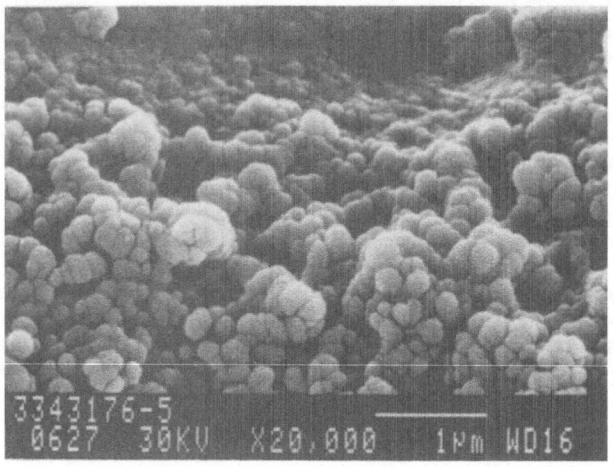

Fig. 7   SEM Photomicrograph of the Outer Edge of an Asymmetric
           Polyethersulfone Hollow Fiber Membrane spun from
           N-methylpyrrolidone/Formamide at 20K Magnification

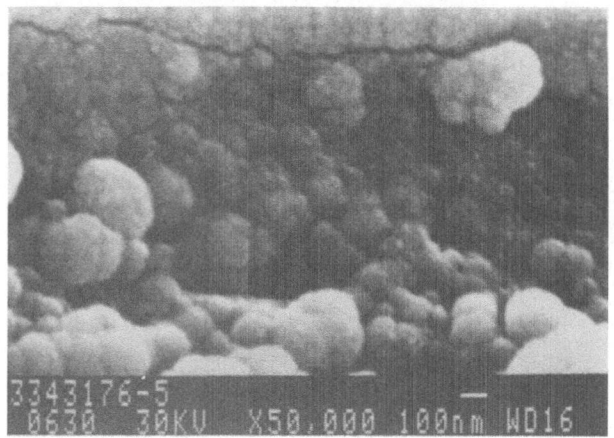

Fig. 8   SEM Photomicrograph of the Outer Edge of an Asymmetric
           Polyethersulfone Hollow Fiber Membrane spun from
           N-methylpyrrolidone/Formamide at 50K Magnification

     The oxygen plasma ablation results are given in Table II and
illustrated graphically in Figure 9. The unetched polyethersulfone
asymmetric hollow fiber membrane, like that made from polysulfone in the
same solvent mixture, exhibits a high helium/nitrogen selectivity. This
selectivity and the helium and nitrogen permeation rates from which it is
calculated remain constant over the first 2-3 minutes of oxygen plasma
exposure. This high selectivity indicates a dense separating layer with
relatively few defects. After three minutes of oxygen plasma treatment,
the nitrogen permeability dramatically increases yielding decline in
selectivity. In contrast, the helium permeability undergoes a gradual
increase in rate with additional oxygen plasma exposure. This behavior
indicates that within the first three minutes of oxygen plasma treatment,
the effective layer is removed and the transitional zone containing ever
increasing quantities of micropores is reached. The minuteness of the

effective separating layer is demonstrated by examination of Figure 10 and the comparison of the photomicrographs in Figure 10 with the unetched control in Figure 7. Figure 10 shows the outer edge of an asymmetric polyethersulfone hollow fiber membrane at 20K magnification after one minute (Figure 10a), three minutes (Figure 10b), and five minutes (Figure 10c) of oxygen plasma exposure. Comparison of these micrographs with each other or with the unexposed control reveals no detectable differences among their skin thicknesses or morphologies.

TABLE II

Performance of Polyethersulfone Hollow
Fiber Membranes Spun from
N-methylpyrrolidone/Formamide as
Function of Oxygen Plasma Exposure Time

Etch Time (min)	$P/\ell(He)^a$ $\times 10^6$	$P/\ell(N_2)^a$ $\times 10^6$	α
0	12.1±2.6	.31±.12	40.1±6.5
1	10.9±1.0	.25±.08	46.6±13.1
2	10.9±0.1	.20±.01	54.5±4.6
3	10.7±0.7	.34±.15	36.2±16.0
5	11.5±0.1	.94±.30	13.0±4.3
8	14.9±2.0	1.70±1.13	11.1±4.9

a   All permeability units are $cm^3(STD)/cm^2\text{-sec-cmHg}$

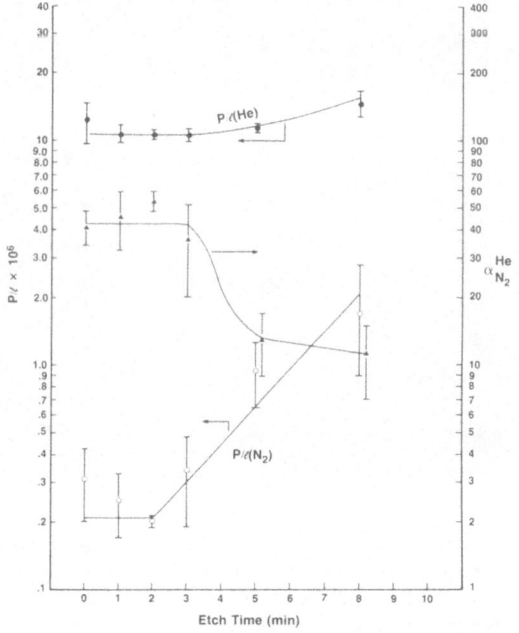

Fig. 9   The Performance of Asymmetric Polyethersulfone Hollow
Fiber Membranes Spun from N-methylpyrrolidone/Formamide
as Function of Oxygen Plasma Ablation Time

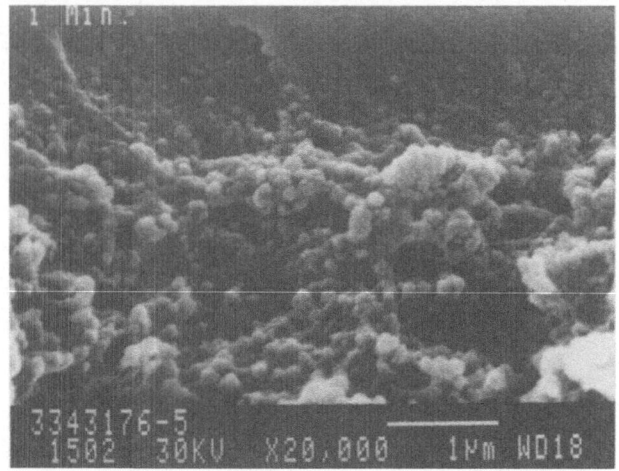

Fig. 10a

Fig. 10b

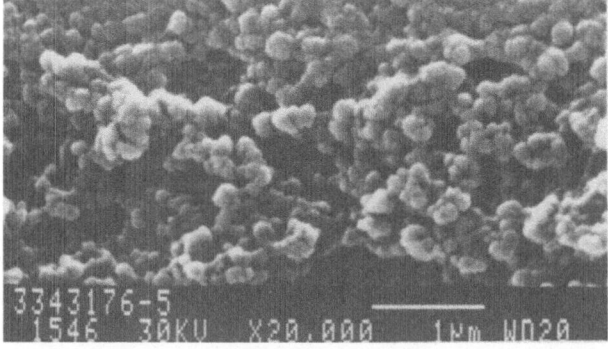

Fig. 10c

Fig. 10   SEM Photomicrographs of the Outer Edge of an Asymmetric
           Polyethersulfone Hollow Fiber Membrane spun from
           N-methylpyrrolidone/Formamide at 20K Magnification
           after Oxygen Plasma Exposure (a) one minute, (b) three
           minutes, and (c) five minutes

Polyphenylsulfone: A scanning electron micrograph of an asymmetric polyphenylsulfone hollow fiber membrane spun from a solution of N-methylpyrrolidone (85 wt%)/formamide (15 wt%) is shown in Figure 11. As with asymmetric polyethersulfone hollow fiber membranes, structural differences in this hollow fiber membrane are seen at both the inner and outer surfaces. However, SEM photomicrographs of the outer edge of this asymmetric hollow fiber cross-section reveal distinct differences between its skin structure, as shown in Figures 12 and 13, and those of asymmetric hollow fiber of polysulfone and polyethersulfone prepared from the same solvent/nonsolvent combination. Examination of these figures reveals that the structure consists also of spherical micelles. The packing density of this micellar array gradually increases with proximity to the outer surface. This increase in packing density is so gradual that discrete pores are observed in the structure between the micelles immediately below the montonous surface layer. This monotonous surface laeyr has the thickness equivalent to the individual micellar diameter (100-200nm) and is composed of these structures. However, in the surface layer the micelles are so tightly packed that their individual structures cannot be resolved.

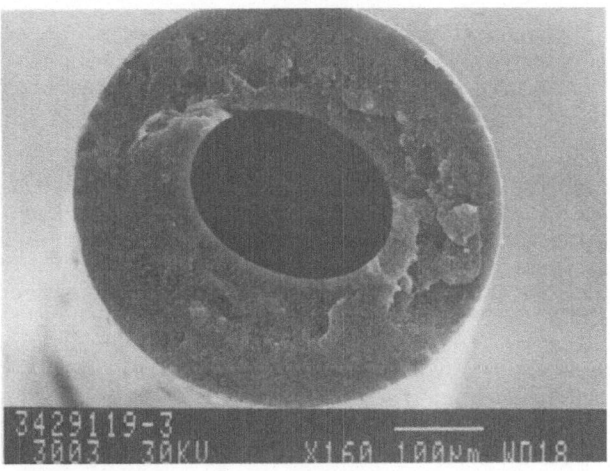

Fig.11   SEM Photomicrograph of a Cross-section of an Asymmetric Polyphenyl-sulphone Hollow Fiber Membrane Spun from N-methylpyrrolidone/Forma-mide at 150 Magnification

Fig.12   SEM Photomicrograph of the outer edge of an Asymmetric Polyphenyl-sulfone Hollow Fiber Membrane Spun from N-methylpyrrolidone/Forma-mide at 20K Magnification

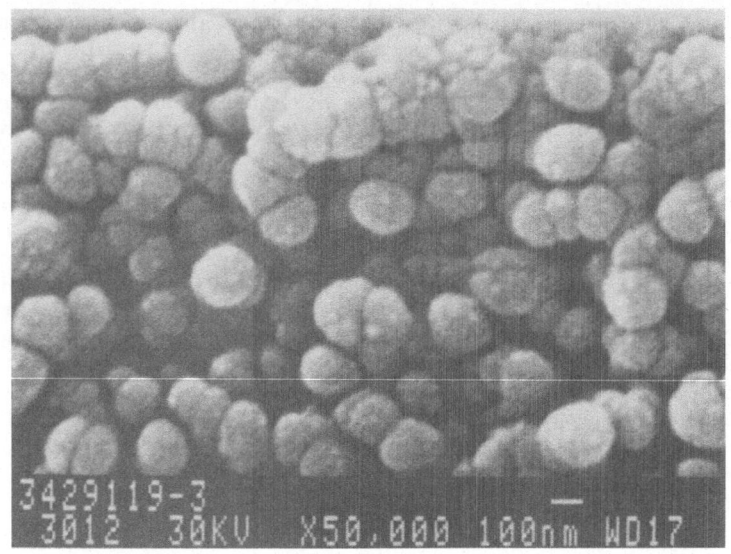

Fig. 13   SEM Photomicrograph of the Outer Edge of an Asymmetric
Polyphenylsulfone Hollow Fiber Membrane spun from
N-methylpyrrolidone/Formamide at 50K Magnification

The results of the oxygen plasma ablation experiments are given in
Table III and shown graphically in Figure 14.   The unetched
polyphenylsulfone asymmetric hollow fiber membrane exhibits high
selectivity.   This selectivity is maintained throughout the first minute of
oxygen plasma ablation, which is indicative of a dense effective separating
layer with a minimum number of defects.   If one can assume that the oxygen
plasma ablation rate is similar for both polysulfone and polyphenylsulfone,
then this effective separating layer is approximately 10-15nm in thickness
or 5-15% of the monotonous surface layer.

TABLE III

Performance of Polyphenylsulfone Hollow
Fiber Membranes Spun from
N-methylpyrrolidone/Formamide as Function
of Oxygen Plasma Exposure Time

Etch Time (min)	$P/\ell(He)^a$ $x10^6$	$P/\ell(N_2)^a$ $x10^6$	α
0	27.2±.5	.83±.28	35.3±10.5
1	17.1	.47	36.5
2	27.2±3.3	2.6±1.1	11.3±3.5
3	40.0±7.3	8.3±3.8	5.2±1.2
5	58.4±2.6	17.9±1.0	3.2±0.1
8	99.4±2.0	38.3±8.7	2.6±0.1

a   All permeability units are $cm^3(STD)/cm^2$-sec-cmHg

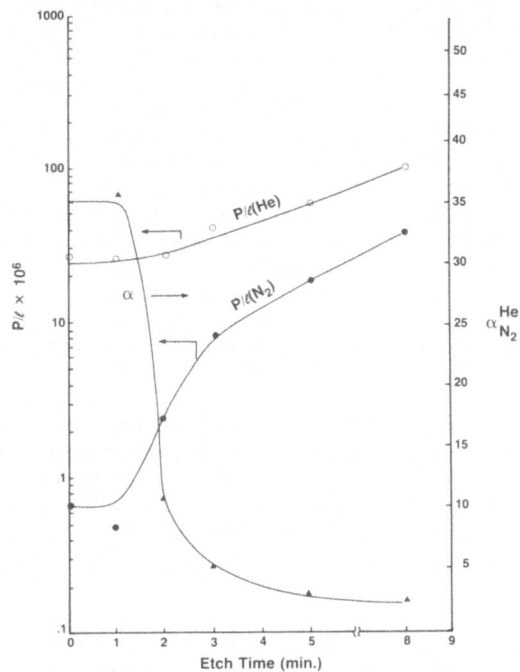

Fig. 14   The Performance of Asymmetric Polysulfone Hollow Fiber
Membranes Spun from N-methylpyrrolidone/Formamide as
Function of Oxygen Plasma Ablation Time

A subsequent decline in the separation factor resulting from a rapid
increase in nitrogen permeability is then seen within the one to four
minute exposure interval. This decline in selectivity occurs as a result
of encountering micropores within the monotonous surface layer below the
effective separating layer. The presence of both the effective separating
layer and the transition zone in a thickness equivalent to that of the
diameter of a microscopically observable micelle implies that these
micelles are in reality aggregations of yet smaller structures.

Within eight minutes of oxygen plasma ablation, the selectivity has
fallen to 2.6, which is equivalent to the separation factor for helium and
nitrogen with Knudsen flow, i.e., a selectivity equal to 2.65.   Knudsen
flow prevails when the pores are substantially smaller than the
mean-free-paths of the diffusing species. At 100 psig and 20°C, the
mean-free-paths of helium and nitrogen are 35nm and 12nm, respectively.
Therefore, the sizes of the micropores in the transitional layer must be
significantly less and may either approach and/or exceed the limits of
resolution of the scanning electron microscope (4nm).

CONCLUSIONS

Asymmetric hollow fiber membranes of polysulfone, polyethersulfone,
and polyphenylsulfone can be prepared by phase inversion spinning
solvent/nonsolvent dopes, i.e., N-methylpyrrolidone/formamide.   These
asymmetric hollow fiber membranes possess a microscopically observable skin
supported by a porous open cellular network. The walls of the open cells
of the matrix are composed of arrays of interconnected spherical micelles.
With increasing proximity to the outer surface, the packing density of the
spherical micelles increase with a concomitant decline in interstitial
porosity. At the outer surface layers, the packing of the micelles becomes

so dense that their boundaries can not be discerned yielding a monotonous surface layer. The thickness of this monotonous surface layer varied from 100-200nm in the asymmetric hollow fiber membrane prepared from polyphenylsulfone to 1200-1500nm for the polyethersulfone samples.

Oxygen plasma ablation is a useful technique with which to investigate the structure of these outer layers. The results of these studies suggest that the dense outer layer is not homogeneous. Instead it is composed of a dense effective separating layer with few penetrating defects, which is only a small fraction of the dense outer layer thickness. Beneath the effective separating layer is a transition region containing micropores at or below the limits of resolution of the scanning electron microscope. Exposure of these micropores by oxygen plasma ablation results in a dramatic decline in selectivity due to rapid increases in the slow gas permeability.

Such behavior was observed in asymmetric hollow fiber membranes prepared from polyphenylsulfone in which the thickness of the monotonous layer is equivalent to that of the diameter of a spherical micelle. Therefore, these results also indicate that the microscopically observable micelles are themselves aggregates of smaller structures.

REFERENCES

1. S. Loeb and S. Sourirajan, U.S. Patent 3,133,132 (1964).
2. R.E. Kesting, "Synthetic Polymeric Membranes, A Structural Perspective", 2nd ed., John Wiley and Sons, New York (1985).
3. J.M.S. Henis and M.K. Tripodi, U.S. Patent 4,230,463 (1980).
4. R.H. Hensen, J.V. Pascale, T. Benedictus, and P.M. Rentzepis, J. Polym. Sc., Part A, 3, 2205 (1965).
5. H. Yasuda, "Plasma Polymerization", Academic Press Inc., New York (1985).
6. E.L. Lawton, J. Appl. Polym Sci., 18, 1557 (1974).
7. R.H. Hansen, in "Interface Conversion", P. Weiss and G.D. Cheevers, Eds., Elseveir, New York (1969).
8. H. Yasuda, C.E. Lamaze, and K. Sakaoku, J. Appl. Polyn, Sci., 17, 137 (1973).
9. G.A. Byrne and K.C. Brown, J. Soc. Dyers Colour., 88, 113 (1972).
10. M. Hudis, J. Appl. Polym. Sci., 16, 2397 (1972).
11. H. Schonhorn and R.H. Hansen, J. Appl. Polym. Sci., 11, 1461 (1967).
12. J.R. Hall, C.A.L. Westerdahl, A.T. Devine, and M.J. Bodnar, "Surface Treatment of Polymers with Activated Gas Plasma for Adhesive Bonding", Technical Report No. 3788, Picatinny Arsenal, Dover, NJ, January 1969.
13. J.R. Hall, C.A.L. Westerdahl, M.J. Bodnar, and D.W. Levi, J. Appl. Polym. Sci., 16, 1465 (1972).
14. H. Schonhorn, F.W. Ryan, and R.H. Hansen J. Adhesion, 2, 92 (1970).
15. N.J. Delollis, Rubber Chem. Technol., 46, 549 (1973).
16. R.S. Sowell, N.J. DeLollis, H.J. Gregory, and O. Montoya, J. Adhesion, 4, 15 (1972).
17. Y. Kitazaki and T. Hata, J. Adhesion, 4 123 (1972).
18. F.R. Anderson and V.F. Holland, J. Appl. Phys., 31, 1516 (1960).
19. F.R. Anderson, J. Appl. Phys., 34, 2371 (1963).
20. E.H. Harris and J.H. Magil, Polymer, 3, 252 (1962).
21. M.R. Padhye, N.V. Bhat, and P.K. Mittal, Textile Res. J., 46, 502 (1976).
22. L.I. Bezruk and Y.S. Lipatov, "International Symposium on Macromolecules, Leiden, The Netherlands, IUPAC", 2, 823 (1970).
23. A.K. Fritzsche, Proceedings of the ACS Division of Polymeric Materials:Science and Engineering, 56, 41 (1987).

RECENT ADVANCES IN SELF DOPED CONDUCTING POLYMERS AND ARYLENEVINYLENES

F. Wudl, A. O. Patil and Y. Ikenoue

Institute for Polymers and Organic Solids
Department of Physics and Department of Chemistry
University of California
Santa Barbara, CA. 93106

INTRODUCTION

The theme of this Symposium is "Functional Polymers". Conducting polymers (reviewed more generally by Professor Wnek in this Symposium) can be viewed as functional polymers; the whole backbone is essentially the polymer's functional group because it is a delocalized π system which undergoes REDOX reactions, also known as *doping* reactions. In the doping process, when the backbone is partially oxidized, the resulting delocalized positive charges need to be balanced by introduction of an unreactive anion from the polymer's exterior; when the polymer is partially reduced, the resulting delocalized negative charges are balanced by insertion of a cation into the polymer's bulk. Concomitant with the REDOX changes are changes in the absorption spectrum of the polymer. The pristine polymers have a bandgap that ranges from 1 - 2.5 eV corresponding to λmax of ~800 nm - ~400 nm; as the polymer is either oxidized or reduced, the oscillator strength of the pristine band diminishes and a new band (or bands) at lower energies (longer wavelengths) appear(s). Since REDOX of the backbone can be performed electrochemically, the spectroscopic changes described above are those of an *electrochromic* material. The speed of reversible electrochromic color switching from, e.g., red to blue and back to red (polythiophene oxidation) is dependent on the rate at which ions can be transported through the bulk of the polymer (see Professor C. R. Martin's contribution in this volume with respect to polypyrrole).

The relatively recent discovery that addition of rather bulky long hydrocarbon chains to the backbone of polythiophenes [to produce processable poly(alkylthiophenes) (PAT's)] had only a minor effect on their conductivity, was counter to the general consensus that even a minor disturbance of interchain interactions between polymer molecules in the solid state would practically obliterate electronic transport. This discovery opened the way to further functionalize conducting polymers by covalent bonding of a typical heteroatomic functional group to the backbone through a hydrocarbon spacer or tether.

In this presentation we describe our work on "self doped polymers" (SDP) which were a natural evolution from the PAT discovery and more recent work on poly(arylenevinylenes), a series of polymers obtainable in very high molecular weight and high yield through a water soluble precursor polymer.

Self doped polymers are conducting polymers which carry a dopant counterion covalently bound to every monomer unit or to any number of monomer units. The principle of SDP is depicted in Scheme I, below:

Scheme I

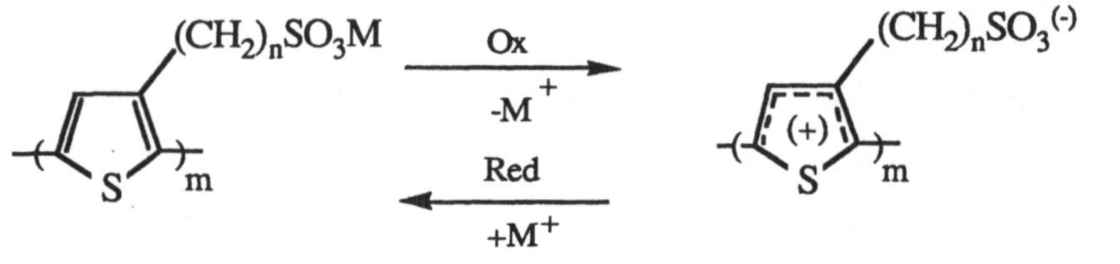

M = H, Li, Na, etc

n = 2, 4

as shown above, the counter-ion is covalently bonded to the conjugated chain; charge transfer is accompanied by proton or sodium ion ejection ("H[+] popping" or "Na[+] popping", respectively) leaving behind the oppositely charged counterion. As examples of self-doped polymers, we reported the successful synthesis and initial characterization of two novel systems[1], sodium salts and acid form of poly 3-(2-ethanesulfonate) thiophene (P3-ETS) and poly 3-(4-butanesulfonate) thiophene (P3-BTS). These are the first known examples[2] of water-soluble conducting polymers. The principal fundamental interest in these polymers rests with the hypothesis that they can lose a proton or other monovalent cation concomitant with electron loss (oxidation) to produce self-doped polymers. The potential counterions are covalently bound via side chains to the polymer backbone; electron ejection from the polymer π-system (positive charging of the backbone) is compensated by proton (or Li[+], Na[+], etc.) migration, leaving behind a covalently bound anion. It is clear that the driving force behind this research is improvement of the electrochromic switching time and an enhancement in the charge storage capacity of polymer-based batteries. Preliminary *in situ* doping experiments during electronic spectroscopy[1] reveal, however, that both the sodium salt and the acid form of the SDP polymers cannot be fully doped.

In this presentation, we review results[3] which directly confirm the validity of the proposed SDP mechanism. By using cyclic voltammetry simultaneously with the determination of hydrogen ion or sodium ion concentration in the electrolyte, we proved that H[+]-ejection and Na[+]-ejection occur upon electrochemical oxidation of the polymer. The results demonstrate, in addition, a voltage controlled ion exchange mechanism.

Cyclic Voltammetry on the Sodium Salt and the Self-Doping Mechanism

Cyclic voltammetric data for polythiophene and some 3-substituted polythiophenes are well-known[4,5]. A typical cyclic voltammogram of a P3-BTS Na film electrode in a solution of 0.1N Bu$_4$NClO$_4$ in acetonitrile was easily recorded[3]. The oxidation potential of a P3-BTSNa film electrode eventually settles near 0.75 V vs. Ag/AgCl; however, this value increases with the water content of acetonitrile electrolyte solutions. This effect of water may have to do with solvation of the sodium ions near the polymer electrode surface. This would explain the increase in current and "reversibility" of the observed waves.

The above cyclic voltammogram revealed[3] a qualitative correlation between potential and color change, indicating that a P3-BTSNa thin film (~ 5 μm) electrode is electrochromic with clearly observable reversible color change between orange and blue on the cathodic reduction and the anodic oxidation reactions, respectively. The color change was particularly evident when a small amount of water was added. These results demonstrate that P3-BTSNa, a polyelectrolyte is also an electroactive material[6].

Cyclic Voltammetry on the Acid Form

A cyclic voltammogram of P3-BTSH film cast on Pt as an electrode in nonaqueous medium of 0.1N-Bu$_4$NClO$_4$-acetonitrile was reported previously[3]. The anodic and cathodic peak potentials were found to be almost identical to those of the sodium salt. However, the film was not stable on extending the cycle sweeps anodically (under the above conditions) beyond ca.1.5 V vs. Ag/AgCl. The kind of water effect described above for the sodium salt was not observed for the acid; the polymeric acid was extremely hygroscopic and hence the film surface was always coated with a layer of water, as determined by infrared spectroscopy. In fact, when the acid film was fully dehydrated, it turned greenish blue with a concomitant loss in conductivity![7]; exposure to water vapor regenerated the original, orange color. We ascribed this aqueous solvatochromism to an acid⇌base reaction where, when the water is completely removed, the next best base to solvate the protons of the sulfonic acid side chains is the polythiophene backbone, as shown schematically below.[8]

H$^+$Ejection Study of the Acid Form of Polymers and the Self-Doping Mechanism

In order to investigate the kinetics of H$^+$ ejection during cyclic voltammetry in close proximity to the W.E. of P3-BTSH, we employed a pH electrode monitor sensitive to the H$^+$ concentration of the nonaqueous solution. A "pH" change in solution was dynamically observed during electrochemical cycling. Although the expected change in "pH" was observed (due to the release or absorption of H$^+$ from the polymer thin film according to the concept described in Scheme 1)[3] between anodic oxidation and cathodic reduction at a slow sweep rate (2mV/sec), the "pH" cycling behavior was extendable to only four cycles. We also observed that before the potential sweep had reached its most cathodic point, the "pH" had relatively rapidly decreased. This behavior is explained as follows: although the 0.1N Bu$_4$NClO$_4$ acetonitrile electrolyte showed a "pH" value of 5.6-5.7, after insertion of a P3-BTSH W.E. into the cell, the "pH" value of the electrolyte solution decreased to about 1 (or sometimes even close to 0). Since P3-BTSH should act as a strong acid, it is reasonable to expect that an irreversible cation exchange between H^+ and Bu_4N^+ may occur spontaneously in acetonitrile solution . Hence, some fraction (α) of the acid functional groups of the polymer would first exchange to form the Bu_4N^+ salt with an associated spontaneous release of H^+, thereby causing a decrease in "pH" and formation of a mixed fraction of the Bu_4N^+ salt and the H^+ form before the initiation of electrochemical potential cycling, as indicated in equation < 1 >.

$$SDP\text{-}SO_3H + Bu_4N^+ \rightleftharpoons \alpha SDP\text{-}SO_3^-Bu_4N^+ + (1-\alpha)SDP\text{-}SO_3H + \alpha H^+ \qquad < 1 >$$

$$0 < \alpha < 1$$

During the electrochemical cycle, two competitive reactions can take place on the same chain:

$$(1-\alpha)SDP\text{-}SO_3H \quad \rightleftharpoons \quad (1-\alpha)SDP^{n+}\text{-}(SO_3^-)_n + \quad n(1-\alpha)H^+ \qquad < 2 >$$

$$\alpha SDP\text{-}SO_3^-Bu_4N^+ \quad \rightleftharpoons \quad \alpha SDP^{m+}\text{-}(SO_3^-)_m \quad + \quad m\alpha Bu_4N^+ \qquad < 3 >$$

If the electrochemical reaction < 2 > were preferred over < 3 >, then a large change in "pH" would be expected. In fact that is precisely what was observed[3] initially following a change in potential. Thus the changes in "pH" generated during oxidation and subsequent reduction are continuously compensated by the spontaneous ion exchange. Because the diffusion of H^+ to the interface of the polymer electrode and electrolyte solution is the rate-determining step when the solution is not stirred, only a small change in "pH" can be observed when the sweep rate is increased to 50 mV/sec. Finally, we note that during a cathodic sweep, the maximum in "pH" appears at ca. 0.65 V vs. Ag/AgCl. This implies that the ion exchange in eq.<1> from H^+ to Bu_4N^+ is much faster than the H^+ absorption rate (reverse reaction in eq. <2>) at potentials lower than 0.65 V (corresponding to the same reduction peak potential, 0.65 V, of the cyclic voltammogram). Of course, in this set of experiments, the usual doping and undoping process concomitant with anion migration from the electrolyte solution (equations <4> and <5>), although unlikely, is invisible.

$$SDP\text{-}SO_3H + ClO_4^- \rightleftharpoons SDP^{N+}\text{-}SO_3H(ClO_4^-)_N \qquad < 4 >$$
$$SDP\text{-}SO_3^-Bu_4N^+ + ClO_4^- \rightleftharpoons SDP^{M+}\text{-}SO_3Bu_4N(ClO_4^-)_M \qquad < 5 >$$

The "pH" changes described above are reasonable; the "pH" range in acetonitrile solvent vs. water is known[9,10,11] and is rather wide. This is attributed to the dielectric constant of the solvent and to the scale of "Donor-Acceptor Numbers" proposed by Gutmann[12,13] (the window of the pH-glass electrode in nonaqueous solvents is ca. -10.2 to more than 30 based on water[14,15]).

In a somewhat similar experiment, Okano[16] et al. reported pH changes at the counter electrode during water electrolysis at a polypyrrole (PPy)[17] coated electrode in aqueous solution. If indeed there was a trace of water on the polymer surface of the P3-BTSH, its electrolysis could produce "pH" changes not related to the proton ejection mechanism. We therefore carried out control experiment and used a PPy electrode (instead of the P3-BTSH film) in the non-aquous solution (0.1N Bu_4NClO_4 in acetonitrile). We found no detectable change in "pH".

In conclusion, we found that "H^+-popping" took place during cyclic voltammetry in the form of H^+ migration through the polymer in spite of the huge excess of Bu_4N^+ in the electrolyte.

Although the H$^+$ ejection behavior obtained in the above experiment constitutes a proof of the proposed mechanism, we observed Na$^+$ ejection in parallel studies; Na$^+$ was released from the P3-BTSNa film electrode upon oxidation during cyclic voltammetry in nonaqueous solvent (0.1N-Bu$_4$NClO$_4$ in acetonitrile)[3]. In a control experiment, we found that there was no detectable concentration of Na$^+$ in the electrolyte after immersing the polymer film in the solvent for 8 hours. Thus, spontaneous ion-exchange against the Bu$_4$N$^+$ of the supporting electrolyte does not occur. Therefore, the reported change in the Na$^+$ concentration released only during the oxidation of the polymer, is indicative of Na$^+$ ejection.

ADVANCES IN ARYLENEVINYLENE POLYMERS

We found several reasons, enumerated below, to investigate the general field of poly(arylenevinylenes):

• Of all the hydrocarbon-based conducting polymers, poly(phenylenevinylene) (PPV) has the highest stability in the undoped state[18].

• PPV can be obtained in molecular weights on the order of 10^5 via a water soluble precursor polymer[18].

• In a process similar to the "Durham polyacetylene", PPV can be stretch-oriented to relatively high molecular alignment[18].

• The yield of the precursor polymer is very high.

The procedure to obtain oriented PPV strips consists of casting a non-oriented, free standing film of a precursor polymer[19] from water and stretch it while heating to the conversion temperature. It occurred to us that an even higher degree of orientation should be achievable if the precursor polymer itself is oriented first, prior to heat treatment. The first method that we tried to orient the precursor polymer was to spin an aqueous gel into a non solvent which would at the same time be a dehydrating medium. In the past, the method of gel spinning has afforded fibers with a very high degree of orientation, including polyacetylene[20]. The precipitation solvent which gave moderate results, at best, was acetonitrile but it was not fast enough in the dehydration step. We then decided to use ion exchange precipitation as a coagulation technique. A standard technique employed to separate large organic cations from aqueous solution of other inorganic ions, involves addition of a large, complex anion such as fluoroborate, hexafluorophosphate, etc to cause precipitation. Indeed, when an aqueous gel of the chloride precursor polymer was spun into a concentrated solution of sodium fluoroborate in water, a white fiber which could be taken up in a spool was formed. The fiber retained some water which allowed a degree of stretch-orientation before additional stretching during conversion to PPV. Alas, the conversion to PPV using the fluoroborate salt was very sluggish, probably due to the low basicity of the fluoroborate anion compared to the chloride anion. With this technique we were able to obtain PPV fibers with a degree of orientation as determined by X-ray scattering of:

$$f_{x\,ray} \frac{<3\cos^2\theta-1>}{2} \cong .955$$

which is on the same order, or slightly higher than previously reported[18].

Organic solvent soluble polyelectrolyte polymers and Incipient Doping

When a viscous solution of the polyelectrolyte precursor polymer was mixed with excess NaX (X= BF_4^-, AsF_6^-, I^-, PF_6^-) in water, a solid precipitated out immediately. These new precursor polymers, poly(p-xylylene-α-dimethylsulfonium salts), after filtering and washing with water and drying, were soluble in polar aprotic solvents such as CH_3CN, DMF etc. The anion exchange procedure to prepare these novel organic solvent soluble polymers was found to be very efficient. Elemental analysis of anion exchanged polymer suggests complete reaction in two hours (BF_4^-) and more than 90% reaction in the case of AsF_6^-. The chloride form of the polyelectrolyte *need not be in solution for rapid exchange to be observed!* This could be shown by IR spectroscopy of a free standing film of poly(p-xylylene-α-dimethylsulfonium chloride), and a spectrum after anion exchange of the same film by dipping the film in aqueous $NaAsF_6$, in the latter case very intense bands due to the hexafluoroarsenate anion could be seen.

Differential scanning calorimetry (DSC) (heating rate of 20° C/min) of precursor polymer (BF_4^- form) was similar to that of precursor polymer (Cl^- form) showing three endothermic peaks at 115°, 176° and 251° C. The sample after DSC was yellow and was partially converted to PPV. The same analysis (20° C/min) of the AsF_6^- precursor polymer, however, showed multiple endothermic peaks between 80° to 199° C. After the DSC scan, the sample was observed to be black and conducting. Two probe compressed powder conductivity measurements yielded conductivity values of 1-3 $\times 10^{-4}$ S/cm. (a four-probe measurement showed a conductivity of 1×10^{-1} Scm^{-1}. An IR spectrum showed reduced intensity of the peaks due to dimethyl sulfide groups but strong peaks due to AsF_6. This result indicates that after the DSC scan, the sample was partially converted and partially doped. We term this new development Incipient Doping. Among other precursor polymers, the PF_6^- and I^- forms also showed incipient doping. A plausible explanation is given in the Scheme below.

Scheme II

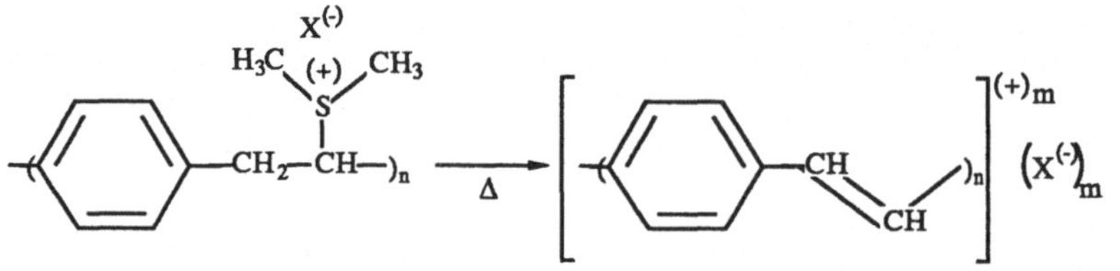

$$X^{(-)} = AsF_6^{(-)}$$
$$X^{(-)} = PF_6^{(-)}$$
$$X^{(-)} = I^{(-)}$$

$$AsF_6^{(-)} \xrightarrow{\Delta} AsF_5 + F^{(-)}$$

$$F^{(-)} + [(ArCH_2CH(Me_2S)] \longrightarrow HF + PPV$$

$$PPV + AsF_5 \longrightarrow [PPV]^{(+)}_m \left(AsF_6^{(-)}\right)_m$$

One piece of evidence which supports the above mechanism is that in order to observe incipient doping, the conversion must be carried out in a very small, closed vessel; i.e., in a large or open container, AsF_5, PF_5, I_2, etc. escape before they have an opportunity to interact with the "nascent" PPV.

104

CONCLUSION

The experimental evidence presented demonstrates that the self-doped polymers (P3-ETS and P3-BTS) act as cation ejectors during oxidation and cation absorbers during reduction. The proposed doping mechanisms (H^+-ejection and Na^+-ejection, respectively) have thus been confirmed. The acid and sodium forms of these 3-substituted polythipohene derivatives can therefore be considered as potential-dependent proton and sodium exchangers. The usefulness of the potential-dependent proton exchange is currently limited by the normal ion exchange which is slower than the potential-dependent one but still significant. It may be possible to control the amount of normal exchange by variations in electrolyte cation and medium. One surprizing result which needs to be investigated is the low degree of charge storage observed spectroscopically for both the sodium salt and the acid.

We have also shown that poly(phenylenevinylene) polyelectrolyte precursor polymers can be processed into fibers by a novel ion exchange precipitation. Ion exchange of these polymers can also be achieved in a phase transfer process on cast films. Finally we have shown that a new concept of *incipient doping* of PPV-type polymers can be realized when the ion exchanged precursor polymers containing the anions PF_6^-, AsF_6^-, or I^-, are heated in a closed vessel to the temperatures required for dimethyl sulfide elimination.

REFERENCES

1. A. O. Patil, Y. Ikenoue, N. Colaneri, J. Chen, F. Wudl, and A.J. Heeger, *Synth. Met.*, 20, 151 (1987).

2. A. O. Patil, Y. Ikenoue, F. Wudl, A.J. Heeger, *J. Am. Chem. Soc.*, 109, 1858 (1987), (b) E. Havinga, L. van Horsen, W. ten Hoeve, H. Wynberg and E. W. Meijer, *Polymer Bull.* 1987, 18, 277. (c) Tsai, W.; Jang, G. W. and Rajeshwar, K. *J. Chem. Soc. Chem. Commun.* 1987, 1776..

3. Some results submitted, Ikenoue, Y.; Chiang, J.; Patil, A. O.; Wudl, F. and Heeger, A. J. *J. Am. Chem. Soc.* in press.

4. G. Tourillon, and F. Garnier, *J. Electroanal. Chem.*, 161:51 (1984).

5. R.J. Waltman, j. Bargon, and A.F. Diaz, *J. Phys. Chem.*, 87:1459 (1983).

6. That this color change is not due to a simple equilibrium:
   $$P3\text{-}BTSNa + H_2O \rightleftharpoons P3\text{-}BTSH$$
   was demonstrated by the fact that the sodium salt changes color reversibly upon chemical oxidation (Br_2 gas) and reduction (hydrazine).

7. Y. Ikenoue, A. O. Patil, unpublished results.

8. A qualitative test for aromatic and heteroaromatic compounds is their solubility in concentrated sulfuric acid (Shriner, R. L.; Fuson, R. C. and Curtin, D. Y. *The Systematic Identification of Organic Compounds*; Wiley: New York, 1956. Another precedent is the H-D exchange of thiophene by dissolving the latter in concentrated D_2SO_4.

9. C. Queck, K. Schwabe, *Meas. Proc. Sci. Technol. Proc. Imeko Congr. Int. Meas. Confed 8 th* , (pub.), 2 , 529-34 (1979).

10. T. Mussini, A.K.Covington,*Chim. Ind. (Milan)* 62(4), 329-32 (1980).

11. *CA.* 12221f. vol.103, 1985 ; "Determination of pH and pA of some non-aquous solvents and their mixtures with inert solvents. *Deposited Doc,* VINITI 3450-84, 11P (1984)".

12. V. Gutmann, "The Donor-Acceptor Approach to Molecular Interactions", Plenum Press 1978

13. V. Gutmann, *Electrochim. Acta* 21, 661 (1976).

14. A. J. Bard, (ed.), *Electroanal. chem.*, vol.8, D.Baner and M.Breant, Solute Behavior in Solvents and Metals, a Study by Use of Transfer Activity Coefficients, P.281 Marcel Dekker 1985.

15. I. M. Kolthoff and P.J. Elving, (ed.), "Treatise on Analytical Chemistry," 2nd ed., Part I. vol. 2 (1979).

16. M. Okano, A. Fujishima K. Honda, *J. Electroanal. Chem.,* **185,** 393 (1985).

17. S. Asavapiriyanont, G.K. Chandler, G.A. Gunawardena and D. Pletcher, *J. Electroanal. Chem.,* **177,** 229 (1984).

18. D. R. Gagnon, F. E Karasz, E. L. Thomas and R. W. Lenz, *Am. Chem. Soc. Polym. Prepr.,* **25,** 284 (1984)

19. R. A. Wessling and R. G. Zimmerman U.S. Pat. 3 401 152 (1968), U.S. Pat. 3 706 677 (1972).

20. J. C. Chiang, P. Smith, A. J. Heeger and F. Wudl, *Polymer,* submitted for publication. (1988).

CHARGE CARRIER CHEMISTRY IN ELECTROACTIVE POLYMERS

Gary E. Wnek

Department of Chemistry
Rensselaer Polytechnic Institute
Troy, NY 12180-3590

INTRODUCTION

It has often been said in casual conversation that the field of conductive polymers is 'materials limited,' meaning simply that existing materials fail to meet one or more of the criteria deemed necessary for a given technological application. Such criteria may include low raw materials and production costs in order to be competitive with existing materials, reasonable processability and mechanical integrity, and perhaps semi-transparency and environmental stability. All are of course intimately linked to polymer chemistry, which is the science of the preparation and molecular-level characterization of macromolecules. The past several years have witnessed important accomplishments in the synthesis of conducting polymers and in the understanding of their molecular structures, and provide the basis for increasing our understanding of key issues such as conduction mechanisms and relationships between structure and physical properties.

In spite of this recent progress mentioned above, a great deal more needs to be learned about molecular structure in particular if we are to seriously exploit applications of these materials. A long-standing problem in molecular characterization (e.g., determination of molecular weights and their distributions, extents of branching (if any), etc.) has been the intractability of most conducting polymers, with crosslinking (the connecting of polymer chains through covalent bonds) typically taking the blame for this unhappy circumstance although there is little (if any) direct evidence for crosslinking. However, the organic backbones of conductive polymers such as polypyrrole are anticipated to be rather stiff and it is perhaps more likely that intractability is the result of low entropies of dissolution and/or melting, with strong π-π intermolecular interactions rendering the free energy of dissolution and/or melting even more unfavorable through the enthalpic contribution. Fortunately, if most conducting polymers are in fact not crosslinked but are intractable for the reasons noted above, it should be possible to mitigate this problem through modifications of the polymer backbones, although in most cases it is desirable that such modifications do not substantially alter electrical properties. Indeed, It is now possible to prepare polymers which are soluble in both neutral and oxidized (conductive) forms.[1]

Another serious problem is the environmental instability of many conductive polymers. Such instability (as reflected, for example, by a decrease of conductivity in air) is not too surprising in view of the fact that the charge carriers are to a chemist modestly delocalized carbenium ions or carbanions in oxidized or reduced derivatives, respectively, and neither ions are noted for their stability towards water and/or oxygen. However, selected conducting polymers such as polypyrrole exhibit remarkably good stability which is probably due to a combination of a low oxidation potential[2] (implying that more stable carbenium ions are formed upon oxidation) and limited access of O_2 and/or H_2O in the 'doped', intercalated structure which may have at least partial crystallinity. Conductive polymers do not have the luxury of naturally forming passivating oxide layers as is the case with traditional semiconductors such as silicon, and hence the charge carriers in the former are in intimate contact with their environment (e.g., an electrolyte in a battery cell). Thus, it is important to understand mechanisms of the reactions of charge carriers in conducting polymers with various reagents if these materials are to be efficiently exploited in technology. As we shall see, charge carriers may be viewed as reactive intermediates for the surface functionalization of conductive polymers and the synthesis of derivatives with dramatically different bulk properties.

COMPARISONS OF CHARGE CARRIER GENERATION AND STABILITY WITH IONIC POLYMERIZATION OF VINYL MONOMERS

A material such as polyacetylene becomes conductive upon 'doping,' meaning in this context the oxidation or reduction of the π-system and incorporation of a charge-balancing counterion. Many similarities exist between doping and ionic chain growth polymerization,[3,4] particularly with the initiation and termination steps. Consider the preparation of p-type (oxidized) polyacetylene, wherein the charge carriers are carbenium ions. Such a material is typically obtained by direct oxidation of the polyacetylene π-system, and the resulting radical cations at sufficiently high concentration will couple, leaving a dication (commonly referred to as a bipolaron).[5] However, protonation[6] (using anhydrous HF) and methylation[7] (using trimethyloxonium hexachloroantimonate) have both been claimed to afford p-type polyacetylene. Thus, oxidative 'doping' of polyacetylene might be generalized as the reaction of polyacetylene with carbenium ion generators, and bears a strong resemblance to the initiation step in the cationic polymerization of vinyl monomers. The analogy is strengthened by the observation that many Lewis acids such as BF_3 require proton donors to be effective as initiators for cationic polymerization. A Brønsted acid is generated which is the true initiator. Interestingly, anhydrous BF_3 is an ineffective 'dopant' for polyacetylene, although introduction of a small amount of water leads to a rapid conductivity increase,[8] presumably the result of protonation of polyacetylene by the strong acid $H^+(BF_3OH)^-$. Acids that can simultaneously act as oxidants (such as H_2SO_4) probably 'dope' via oxidation. Sorensen[9,10] found from an NMR study of protonation of polyenes using H_2SO_4 that polyenes containing more than about five double bonds in conjugation suffer oxidation and radical cation formation, as suggested by the broad NMR lines and odor of SO_2.

It should be noted that the preparation of n-type (reduced) polyacetylene using strong organic bases (e.g., alkyl lithium compounds)[11] or more commonly electron transfer reagents (e.g., sodium naphthalide radical anion)[8] employs the two major classes of initiators used in anionic polymerization of monomers such as styrene and butadiene.[3] Reductive 'doping' can also be accomplished by deprotonation of, for example, acetylene/butadiene copolymers[12] and related phenylenepentadienylenes.[13]

Initiation steps in ionic polymerization of olefins, and charge carrier generation in polyacetylene, may appear to encompass two mechanistically distinct reactions, namely single electron transfers and two-electron additions to the π-system involving either electrophiles or nucleophiles. Examples of the latter include alkylation and protonation. However, Pross[14] has recently criticized the validity of such a distinction and has championed the idea that both reactions are described by a common mechanism termed a *single electron shift*. In this context, protonation of polyacetylene could be viewed as a shift of one electron from the polymer to the proton, affording a polyacetylene radical cation and a hydrogen atom, followed by coupling of the two radicals. Oxidation of polyacetylene using $NO^+PF_6^-$ could also be described as a single electron shift, although here there is little tendency for the polyacetylene radical cation to couple with the resulting $NO\cdot$, and the radical cation prevails. Thus, in principle, 'doping' of polyacetylene using oxidants such as $NO^+PF_6^-$ or Brønsted acids could be viewed as occurring via a common reaction pathway, the single electron shift.

While there is no clear analogy between carrier generation and the propagation step in ionic polymerization (except for perhaps crosslinking), analogies with chain termination and transfer steps are quite appropriate. Good environmental stability of a conducting polymer implies charge carrier longevity, in turn suggesting that carriers are virtually free of termination (*but not necessarily chain transfer*) reactions. Kinetic stability afforded as the result of crystallinity is important. However, consideration should be given to factors which prove to be important in ionic polymerizations as well. For example, in the cases of p-type polyacetylene and cationic polymerization, a gegenion with a low nucleophilicity should be selected to discourage combination.

It is known that p-type polyacetylene slowly loses conductivity even under vacuum or argon[15] and that n-type polyacetylene, while exceedingly reactive in air, is stable up to <u>ca</u>. 200°C under vacuum or argon.[16] We believe that a comparison with ionic polymerization of a diene such as butadiene is instructive here. Polybutadiene prepared by cationic polymerization is branched and crosslinked,[17] although anionic polymerization affords a linear material, a high degree of control over the isomer (cis vs. trans) content, and a narrow molecular weight distribution. These observations imply that, at least for this specific case, the propagating carbanions are more stable intrinsically than their carbenium ion counterparts. The propensity of delocalized carbenium ions to undergo reactions such as cyclizations and hydride shifts is well documented.[4,9,10] Also, in the case of anionic polymerization, combination of the carbanion with an alkali metal cation is not an issue. Indeed, many more examples of 'living' anionic polymerization of vinyl monomers are known compared with cationic polymerization. It is thus not too surprising that n-type polyacetylene is intrinsically more stable than p-type derivatives and more practical for use as a battery electrode (of course in the absence of oxygen and proton donors).

N-TYPE (REDUCED) POLYACETYLENE AS A REACTIVE INTERMEDIATE

We have been interested in approaches for the preparation of hybrid polymers, in this context meaning blends and graft and block copolymers, containing polyacetylene as the electroactive component. The motivation has been the development of conductive polymers having good physical properties without great sacrifice in conductivity upon 'doping.' It occurred to us that the charge carriers in 'doped' polyacetylene may serve as reactive intermediates for the preparation of novel derivatives. In particular, such an approach could be used for the synthesis of graft copolymers wherein growth of the 'arms' is initiated by the charge carriers.

Initial experiments focused on the ability of n-type polyacetyene to initiate the polymerization of monomers such as styrene, methyl methacrylate and ethylene oxide.[18,19] Polymerization of styrene and MMA is rapid and in some cases the initial polyacetylene film disintegrates and a blue 'solution' is formed (presumably a colloidal dispersion of polyacetylene with grafted chains). Subsequent work with powdered n-type polyacetylene (higher surface area) claimed the formation of a soluble graft copolymer.[20] Ethylene oxide also polymerizes on polyacetylene although more slowly. Interestingly, it has been found that n-type polyacetylene reacted with ethylene oxide can be reduced (n-type 'doped' again) to a higher level (higher carbanion concentration) than a film without such a treatment.[21] This enhancement of charging capacity may be important in battery technology and is presumably due to the formation of a passivating PEO layer on the surface of the polyacetylene whose function is not well understood.

As with any synthesis of graft copolymers, two key parameters to control are the average length of the grafts and the average spacing between grafts. The intractability of polyacetylene renders characterization of graft copolymers difficult, and therefore we decided to attach 'arms' of well defined length to eliminate graft length as a variable. Alkyl halides were chosen as the source of the 'arms' principally because of the extensive literature on their reactions with small molecule radical anions[22] (polarons) and dianions[23] (bipolarons). Key issues of interest to us included the mechanism of the reaction with n-type polyacetylene (in keeping with earlier literature, electron transfer vs. S_N2), the influence of the number of alkyl groups (which should introduce sp^3 defects in the polyene backbone) on electrical conductivity, and the use of this chemistry as a means of functionalizing polyacetylene.

Treatment of n-type polyacetylene with any of a variety of alkyl halides in THF typically leads to a dramatic swelling of the film and the formation of an orange-red, soluble fraction.[24] The amount of alkylation of the insoluble but swellable films corresponds to ca. 0.9-1.3 alkyl groups per 10 CH units. Studies of the amount of alkylation of the polyacetylene and the amount of reduction products (R-H and R-R) formed as a function of alkyl halide type leads us to conclude that the mechanism involves initial electron transfer (perhaps more appropriately an electron shift) to R-X as shown in Scheme I.

The resulting R· can suffer one of two fates, coupling with a polyacetylene radical or subsequent reduction to a carbanion, the latter affording reduction products. The mechanism is identical to that proposed for the reaction of disodium tetraphenylethylene with alkyl halides.[23,25]

Additional support for an electron transfer-like mechanism comes from the observation that reactions with n-alkyl halides are incomplete, with ca. 40% of substrate unreacted, whereas reactions with alkyl iodides are nearly complete. This is consistent with the fact that the reducing power of n-type polyacetylene decreases as the carbanion concentration decreases, and that normal alkyl chlorides are the most difficult to reduce in the series employed in our experiments (see Table 1). (Note that the reduction potentials of polyacetylene and the alkyl halides cannot be compared directly as they were measured in different solvents.) Alkyl fluorides, as expected, do not react to a measurable extent.

Scheme 1

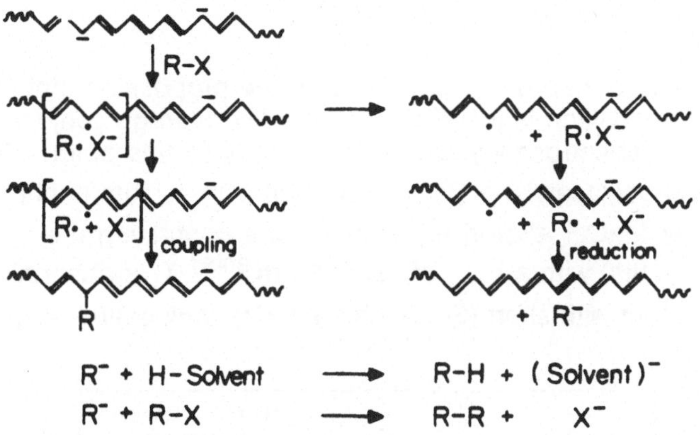

Counterions omitted for clarity. Large brackets indicate cages.

Table 1

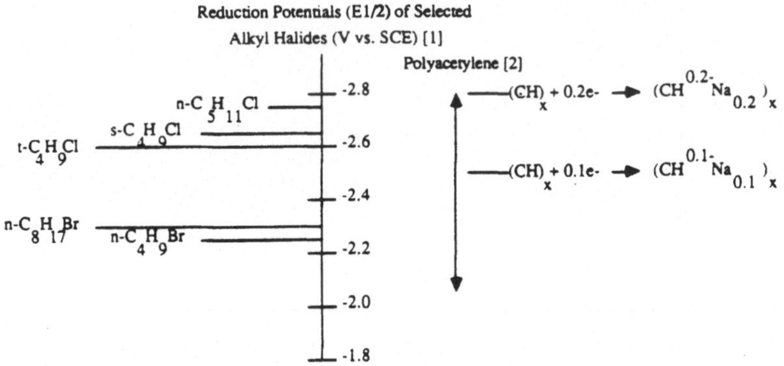

Reduction Potentials ($E1/2$) of Selected
Alkyl Halides (V vs. SCE) [1]

[1] Lambert and Ignall, Tett. Lett., 36, 3231 (1974); Lambert, J. Org. Chem., 31, 4184 (1966)

[2] MacDiarmid et al., "Handbook of Conductive Polymers", T. Skotheim, ed, Dekker, p. 696 (1986)

A remarkable observation is that, while the alkylated films contain ca. 10 % sp³ defects, conductivities upon oxidation with iodine are a respectable 1-5 S/cm.[26] It is interesting that protonation of n-type polyacetylene using methanol affords high levels of defects and yet such a material also yields respectable conductivities upon iodine treatment.[27] This led to the conclusion that rather short conjugation lengths (e.g., tetraenes) are sufficient for reasonable conductivity. Intuitively, conjugation should be important in two respects. First, carrier generation upon oxidation or reduction should be facilitated with high conjugation lengths since the resulting radical cation or anion will be more highly delocalized. Second, higher conjugation lengths ought to facilitate intermolecular charge transport by providing more frequent π-π overlap between adjacent backbones. Indeed, a study[28] of a series of acetylene/methyl acetylene copolymers indicated that conductivity decreases precipitously as the methyl acetylene content exceeded ca. 35 mole%, beyond which the probability of two methylene acetylene units being adjacent (and disrupting conjugation via methyl-methyl steric interactions) becomes more likely.[29] These observations, in addition to the fact that the alkylated and particularly the protonated polyacetylenes are *visually not too different* from the 'undoped' polymer, argue against the idea that short conjugation lengths are responsible for the observed conductivities.

A different picture emerges if one accepts the proposition that knowledge of the sp³ defect composition is insufficient to predict average conjugation lengths. Data from UV-VIS spectroscopy speak specifically to this point. For example, treatment of trans-polyacetylene (λ_{max} = 700nm) with sodium naphthalide radical anion followed by MeOH affords a polyacetylene containing methylene defects (ca. 1 per 10 CH units) with a λ_{max} of ca. 620 nm.[30,31] This absorption maximum is clearly inconsistent with short (3-6 double bonds) conjugation lengths. In fact,

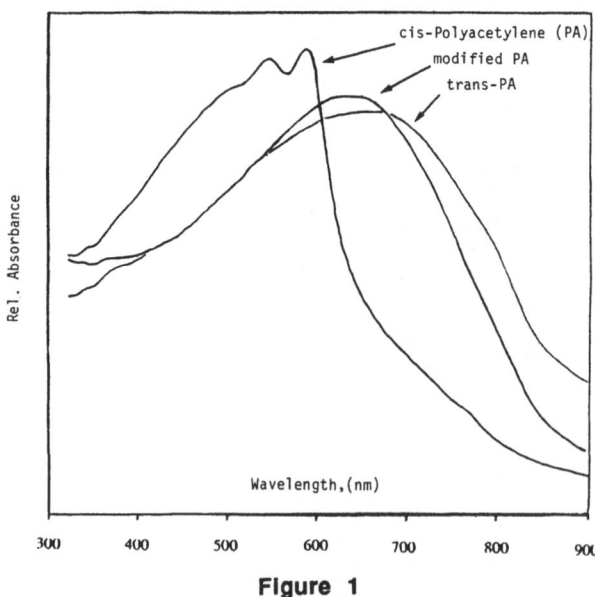

Figure 1

UV-VIS Spectra of cis- and trans-polyacetylene, and 'modified' polyacetylene (reduced with sodium naphthalide and alkylated with 1-bromopentane)

short polyene segments evidenced themselves only after the reductive 'doping'/methanol treatment was repeated twice. We have examined the spectrum of a polyacetylene film alkylated (<u>ca</u>. 1 C_5 per 10 CH units) using 1-bromopentane[26] (Figure 1) and find a λ_{max} again near 620 nm. Thus we are forced to conclude that rather long conjugation lengths remain and that the defects (either -CH_2- or -CHR-) are clustered in the polyacetylene. This idea is supported by preliminary X-ray diffraction data which indicate the presence of pristine polyacetylene in alkylated films and ESR spectra whose linewidths are only slightly broader than those of pristine *trans*-polyacetylene.[26] The alternative conclusion, that methylene defects may not seriously interrupt conjugation and thus would be tolerable even if placed randomly, is intuitively untenable.

There are two reasons why protonation or alkylation of n-type polyacetylene is likely to afford sp^3 defects which are not randomly situated on the polymer backbone. The first is that polyacetylene is semi-crystalline, and anticipated faster kinetics of reaction in the 'amorphous' phase would leave unscathed large segments of conjugation. The second, and likely more important, point is based on a thermodynamic argument, namely that conjugated sequences are more stable than conjugation-interrupted sequences. For example, the resonance energy of conjugated dienes obtained from heats of formation is roughly 3-4 kcal/mol. However, even if protonation (say, using methanol) was to occur rapidly and randomly (kinetic control), the base generated upon protonation (e.g., CH_3O^-), or polyacetylenic carbanions on adjacent chains, can catalyze isomerization to the thermodynamic (conjugated) product. Dias and McCarthy[32] demonstrated that polybutadiene can be isomerized to poly(acetylene-co-ethylene), which was found to contain oligoacetylene units of up to 10 double bonds. The key point is that the -CH_2CH_2-'defects' in polybutadiene, in the presence of a base catalyst in this case, tend to cluster, as do the double bonds. It also should be noted that 'doping' with electron acceptors such as I_2 *after* the introduction of the defects presents another opportunity to attain long conjugation lengths through a series of hydride shifts, a well established concept in organic chemistry. Thus, our earlier analogy with ionic chain polymerization can be extended; the clustering of defects is likely the result of what are appropriately termed chain transfer reactions. Recent band structure calculations[33] indicate that the optical properties of protonated or methylated polyacetylene (produced from the n-type polymer) cannot be accounted for by inferring the presence of short conjugation lengths.

The idea of clustering of defects may help to explain why reasonably high conductivities are observed upon 'doping' of polyacetylene with HF or trimethyl oxonium hexachloroantimonate. For example, random protonation could be followed by a series of hydride transfers which allow the system to move toward equilibrium (i.e., toward longer conjugation lengths). An example (unproven) of such an event is illustrated below.

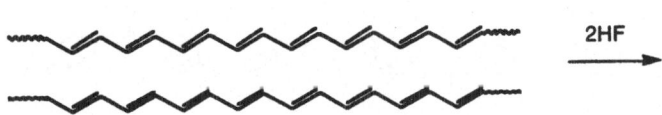

2HF

H Transfer

Much work needs to be done to understand what factors determine the extent of defect clustering, the homogeneity of alkylation or protonation with respect to film thickness, and reaction kinetics. The clustering of defects is a fortunate circumstance in that it allows for introduction of rather large amounts of defects while still preserving reasonable conjugation lengths. It is conceivable that reactions of carbanions in n-type polyacteylene may offer a convenient means of introducing useful (but protected) functional groups for subsequent derivatization. One application may utilize polyacetylene derivatized with chiral groups for asymmetric hydrogenations, wherein catalytically active nickel metal has been deposited on the polymer by treatment of n-type polyacetylene with solutions of $NiBr_2$.[34]

POLYANILINE: A MIXED VALENCE POLYMER WITH MASKED CHARGE CARRIERS

The 'doping' of polyaniline represents another example of interesting chemistry, particularly with regard to charge carrier generation. A few key points are summarized in the following paragraphs.

Several papers between 1908 and 1913 suggested that the principal product of the polymerization of aniline using redox initiating systems is a material known as emeraldine, one of five oxidation states of a linear aniline octamer.[35,36] The para coupling and degree of polymerization of eight were suggested based on wet chemical analyses. The synthesis of the phenyl-capped emeraldine octamer was recently reported[37] and the material exhibits properties (optical, electrical, magnetic; see below) which are similar to those of 'typical' emeraldine (hereafter referred to as polyaniline). It was observed in the 1960's that polyaniline became conducting upon treatment with Brønsted acids such as HCl,[38] and subsequent studies demonstrated a concomitant dramatic increase in paramagnetism upon protonation.[39] The conductive form is stable in acidic, aqueous solutions.

Emeraldine can be considered as a copolymer of p-phenylene diamine and p-quinonediimine units:

The diamines are isoelectronic with hydroquinone (an electron donor) while the diimines are isoelectronic with p-benzoquinone (an electron acceptor). It was recently proposed[40] that this 'built-in' redox couple, which is pH-sensitive, is the source of charge carriers in the form of semiquinone radical cations[41] which are generated upon protonation of the quinonediimine, whose accepting power is expected to increase on protonation. A subsequent perusal of the literature indicates that a similar idea was proposed many years ago.[42] A key issue yet to be resolved is whether electron transfer from diamine to diimine is an inter- or intramolecular process. An alternative mechanism of carrier generation which involves spin unpairing of diimines (creating a pair of radical cations) has recently been proposed.[43] Carrier transport is believed to occur via intermolecular redox reactions between radical cations and diamine and/or diimine units.

The resistivity of polyaniline (pH ≈ 1.5) is dependent upon applied potential. The resistivity exhibits a minimum whose magnitude and position with respect to potential depends upon pH (Figure 2) as was first observed by Paul et al.[44] The electrochemistry of polyaniline has been studied in some detail and provides for an interpretation of the resistivity data.[45-48] The first anodic wave, which corresponds to the oxidation of diamine units to radical cations, is essentially pH-independent. However, the second wave, which is the result of oxidation of radical cations to quinone diimines, is pH-dependent as the result of immediate deprotonation of the quinonediimine. Thus, the second wave moves cathodically (closer to the first wave) as pH increases, and the region of minimum resistivity (Figure 2)[48] in turn narrows while the magnitude of minimum resistivity increases. The resistivity is therefore intimately dependent on two parameters, pH and potential.

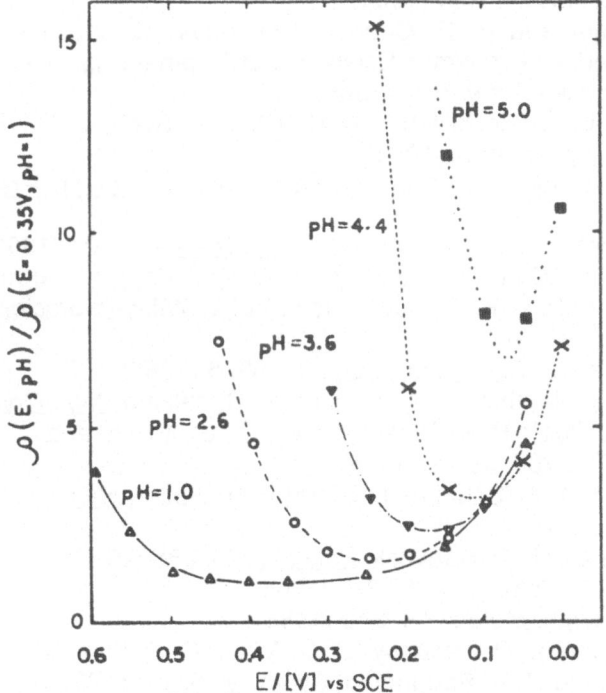

Figure 2
Influence of pH and potential on the resistivity of polyaniline films

Recent studies[49] on poly(o-methylaniline) have shown that the width of the window of minimum resistivity is smaller than for polyaniline, and is very narrow (the first and second redox waves are barely separable) for poly(o-ethylaniline). The first waves are more anodic while the second waves are more cathodic, and we invoke electronic and steric arguments to rationalize these observations. The key point is that judicious selection of substituents may allow for the design of electroactive polyanilines having a unique pH/potential/conductivity response which may find use as microelectronic devices.[44]

ACKNOWLEDGEMENTS

The author wishes to thank members of his research group, in particular G. F. Dandreaux, D. H. Whitney, W. W. Focke and Y. Wei, for their outstanding effort. This work was sponsored by a grant from the Office of Naval Research and a grant from DARPA monitored by ONR.

REFERENCES

1. K. Y. Jen, R. Oboodi, and R. L. Elsenbaumer, Polym. Mat. Sci. Eng., 53:79 (1985); M. A. Sato, S. Tanaka, and K. Kaeriyama, Synth. Metals, 14:289 (1986); S. Hotta, S. D. D. V. Rughooputh, A. J. Heeger, and F. Wudl, Macromolecules, 20:212 (1987).
2. A. F. Diaz, J. Crowley, J. Bargon, G. P. Gardini, and J. B. Torrance, J. Electroanal. Chem., 121:355 (1981).
3. G. Odian, "Principles of Polymerization," 2nd ed., Ch. 5, McGraw-Hill, New York (1981).
4. J. P. Kennedy and E. Maréchal, "Carbocationic Polymerization," Wiley-Interscience, New York (1982).
5. J. E. Frommer and R. R. Chance, Electrically Conductive Polymers, in: Encyclopedia of Polymer Science and Engineering, Vol. 5, 2nd ed., p. 462-507, Wiley, New York (1986).
6. T. C. Clarke, B. W. McQuillan, J. F. Rabolt, J. C. Scott, and G. B. Street, Mol. Cryst. Liq. Cryst., 83:1 (1982).
7. P. Brant, D. Weber, R. Mowery, and R. Nowak, Polym. Prepr., 23(1): 98 (1982).
8. A. J. MacDiarmid and A. J. Heeger, Synth. Metals, 1:101 (1980).
9. T. S. Sorensen, Dienylic and Polyeneylic Cations, in: "Carbonium Ions,"G. A. Olah and P. v. R. Schleyer, eds., Vol 2, Wiley-Interscience, New York (1970).
10. T. S. Sorensen, J. Am. Chem. Soc., 87: 5075 (1965).
11. B. Francois, C. Mathis, R. Nuffer, and A. Rudatikira, Mol. Cryst. Liq. Cryst., 117:113 (1985); R. L. Elsenbaumer, G. G. Miller, and J. E. Toth, U. S. Patent 4 526 708, July 2, 1985.
12. L. M. Tolbert, J. A. Schonmaker, and F. J. Holler, Synth. Metals, 15: 195 (1986).
13. B. Gordon III and L. F. Hancock, Polymer, 28: 585 (1987).
14. A. Pross, Acc. Chem. Res., 18:212 (1985).
15. M. A. Druy, Synth. Metals, 15: 243 (1986).
16. R. L. Elsenbaumer, P. Dellanoy, G. G. Miller, C. E. Forbes, N. S. Murthy, H. Eckhardt, and R. H. Baughman, Synth. Metals, 11: 251 (1985).
17. Ref. 3, Ch 8.
18. G. F. Dandreaux, M. E. Galvin, and G. E. Wnek, Org. Coat. Appl. Polym. Sci., 48: 541 (1983).

19. M. E. Galvin, G. F. Dandreaux, and G. E. Wnek, in: "Polymers in Electronics," T. Davidson, ed., pp. 507-514, ACS, Washington, D. C. (1984).

20. I. Kminek and J. Trekoval, Makromol. Chem. Rapid Commun., 5:53 (1984).

21. M. Maxfield, J. F. Wolf, G. G. Miller, J. E. Frommer, and L. W. Shacklette, J. Electrochem. Soc., 133: 117 (1986).

22. J. F. Garst, Acc. Chem. Res., 4: 400 (1970).

23. J. F. Garst, R. D. Roberts, and J. A. Pacifici, J. Am. Chem. Soc., 99: 3528 (1977).

24. D. H. Whitney and G. E. Wnek, Macromolecules, 21: 266 (1988).

25. M. C. R. Symons, J. Chem. Res., Synop., 360 (1978).

26. D. H. Whitney, Ph.D. Thesis, Dept. of Materials Science and Engineering, MIT (1987).

27. S. I. Yaniger, M. J. Kletter, and A. G. MacDiarmid, Polym. Prepr., 25(2): 264 (1984).

28. J. C. W. Chien, G. E. Wnek, F. E. Karasz, and J. A. Hirsch, Macromolecules, 14: 479 (1981).

29. S. K. Tripathy, D. Kitchen, and M. A. Druy, Macromolecules, 16: 190 (1983).

30. K. Soga, S. Kawakami, H. Shirakawa, and S. Ikeda, Makromol Chem. Rapid. Commun., 1: 643 (1980).

31. X. Q. Yang, D. B. Tanner, G. Arbuckle, A. G. MacDiarmid, and A. J. Epstein, Synth. Metals, 17: 277 (1987).

32. A. J. Dias and T. J. McCarthy, Macromolecules, 18: 869 (1985).

33. S. B. Clough, S. K. Tripathy, X. F. Sun, B. Orchard, and G. E. Wnek, submitted.

34. D. H. Whitney and G. E. Wnek, Mol. Cryst. Liq. Cryst., 121: 313 (1985).

35. R. Willstätter and S. Dorogi, Chem. Ber., 42: 2143 (1909).

36. A. G. Green and A. E. Woodhead, J. Chem. Soc., 97: 2388 (1910); ibid., 101: 1117 (1912).

37. F.-L. Lu, F. Wudl, M. Nowak, and A. J. Heeger, J. Am. Chem. Soc., 108: 8311 (1986).

38. M. Josefowicz, L. T. Yu, J. Perichon, and R. Buvet, J. Polym. Sci. Part C, 22: 1187 (1969).

39. A. G. MacDiarmid, J.-C. Chiang, A. F. Richter, and A. J. Epstein, Synth. Metals, 18: 285 (1986).

40. G. E. Wnek, Synth. Metals, 15: 213 (1986).

41. P. M. McManus, S. C. Yang, and R. J. Cushman, J. Chem. Soc. Chem. Commun., 156 (1985).

42. V. V. Kopylov, A. N. Pravednikov, V. M. Vozzhennikov, and V. K. Bel'skii, Russ. J. Phys. Chem., 52: 305 (1978).

43. F. Wudl, R. O. Angus, Jr., F. L. Lu, P. M. Allemand, D. J. Vachon, M. Nowak, Z. X. Liu, and A. J. Heeger, J. Am. Chem. Soc., 109: 3677 (1987).

44. E. W. Paul, A. J. Ricco, and M. S. Wrighton, J. Phys. Chem., 89: 1441 (1985).

45. W.-S. Huang, B. D. Humphrey, and A. G. MacDiramid, J. Chem. Soc., Faraday Trans. 2, 82: 2385 (1986).

46. P. M. McManus, R. J. Cushman, and S. C. Yang, J. Phys. Chem, 91: 744 (1987).

47. T. Kobayashi, H. Yoneyama, and H. Tamura, J. Electroanal. Chem., 177: 281 (1984).

48. W. W. Focke, G. E. Wnek, and Y. Wei, <u>J. Phys. Chem.</u>, 91: 5813 (1987); W. W. Focke, Ph.D. Thesis, Department of Materials Science and Engineering, MIT (1987).

49. Y. Wei, W. W. Focke, G. E. Wnek, A. Ray, and A. G. MacDiarmid, submitted.

ELECTROCHEMICAL INVESTIGATION OF

ELECTRONICALLY CONDUCTIVE POLYMERS

Charles R. Martin, Reginald M. Penner, and Leon S. Van Dyke

Department of Chemistry
Texas A&M University
College Station, TX 77843-3255

INTRODUCTION

Electronically conductive polymers are a fascinating new class of materials with unique electronic, electrochemical, and optical properties (1). One of the most interesting aspects of these polymers is that they can be reversibly "switched" between electronically insulating and electronically conductive states. Numerous researchers in a variety of different fields of science are currently investigating this insulator/ conductor transition (1). In addition to being an interesting scientific phenomenon, this switching process plays an integral role in nearly all of the proposed chemical applications of electronically conductive polymers (2-13).

Assuming that the polymer is insulating in the reduced form and con- ductive in the oxidized form, the insulator-to-conductor switching reaction could be expressed as:

$$-(Py)_z- \ + \ nX^- \quad ----> \quad -(Py)_{z-n}(Py^+X^-)_{n^-} \ + \ ne^- \qquad [1]$$

where $-Py-$ and $-Py^+-$ symbolize reduced and oxidized monomer units, respectively, and X^- is a charge compensating anion; X^- is initially present in a contacting solution phase but, upon oxidation of the polymer, is incorporated into the polymer phase. Equation 1 shows that, the switching reaction involves a charge-transport process, in which oxidized monomer sites and charge compensating counterions move through the polymer film.

While the case is far from closed, the bulk of the currently available evidence suggests that the rate of this charge-transport process is controlled by ionic diffusion in the polymer phase (14-18). Because diffusional mass-transport plays a role in many electrochemical experiments, it should be possible to assess the rate of the conductor/insulator switching reaction using an electrochemical method. For example, large amplitude electrochemical methods have been used to assess the rates of charge-transport in redox polymer films on electrode surfaces (19-24). These experiments yield an apparent diffusion coefficient, D_{app}, which is a quantitative measure of the charge-transport rate (19-24).

Because large amplitude electrochemical methods have proven useful for evaluations of charge-transport rates in redox polymers(19-24), analogous experiments have recently been conducted on polypyrrole and other electronically conductive polymers (25-29). While this may appear to be a logical extension of the redox polymer work, we have shown that it is difficult, if not impossible, to obtain meaningful D_{app}, data for conducting polymers using large amplitude methods (30).

We have recently developed a small amplitude current-pulse method which is uniquely suited for solving the myriad problems associated with electrochemical analyses of electronically conductive polymers (30, 31). This method was used to obtain D_{app} values for the electronically conductive polymer, polypyrrole (30). The results of these experiments are described in this manuscript.

In addition, we have recently described a procedure for controlling the morphologies of electronically conductive polymers (32). This procedure involves the electrochemical growth of the conductive polymer at an electrode surface which has been masked with a microporous polymer membrane. The pores in this membrane act as templates for the nascent electronically conductive polymer. Because the template membrane contains linear cylindrical pores, cylindrical conductive polymer fibrils are obtained (32). We will show in this manuscript that microfibrillar polypyrrole films prepared via this approach can support higher rates of charge-transport than conventional polypyrrole.

THEORY SECTION

Why Large Amplitude Methods Should not be Used to Obtain D_{app}'s for Electronically Conductive Polymers

Our objections to large amplitude methods can be illustrated by considering a chronoamperometric experiment at a conducting polymer film-coated electrode (33). The first problem which must be addressed concerns the area of the electrode. Analysis of chronoamperometric data via the Cottrell equation requires semi-infinite linear diffusion to a planar electrode of known area (34). If the experiment is initiated in the potential region where the polymer film is conductive, one has, in effect, a large area porous metal electrode (15,31). Thus, the following questions arise - is diffusion in this porous electrode linear and what is the electrode area (35)? These questions are complicated by the fact that during the course of the experiment, the film becomes non-conductive; clearly the electrode area changes, but how and when?

One might suggest that the above criticism could be circumvented by initiating the potential step in a potential region where the polymer is reduced. In this case, at time = 0, the substrate conductor defines the electrode area. Therefore, initially, the area is known and diffusion is linear. However, at some time during the course of the experiment, the film becomes conductive and the unresolved questions, noted above, return.

The second problem concerns the capacitive current (i_c). In the potential region where the polymer is an electronic insulator, the film-coated electrode shows capacitive currents which are similar in magnitude to i_c's at the corresponding uncoated electrode. We have shown that these currents are associated with charging the double layer at the electrode/reduced-film interface (30). In contrast, in the potential region where the polymer is conductive, a much larger "capacitive-like" current is observed. This larger i_c has been ascribed to charging the double layer associated with the conductive polymer film itself (15,30,31,36)

The term "capacitive-like" was used above because, although this current has the qualitative appearance of an i_c, it is clear that this current results from oxidation (or reduction) of the polymer just as the "faradaic" portion of the current (i_f) results from oxidation (or reduction) of the polymer (37,38). The following fundamental question arises - because both the "faradaic" and "capacitive" currents result from the same chemical process, should the apparent i_c be subtracted from the total current to obtain i_f? Furthermore, because the capacitances of the bare and coated electrodes are very different, background correction would be very problematic (30,31).

In the conventional chronoamperometric experiment, it is often possible to circumvent the i_c problem by waiting for i_c to decay (34). However, because the time course of delivery of the large, capacitive-like current at the conductive polymer-coated electrode is unknown, a correction based on temporal discrimination also seems doubtful. Finally, one might suggest that this i_c problem could be circumvented by conducting a chronocoulometric experiment, in that double layer charging is typically observed as an intercept in the Q vs. $t^{1/2}$ plot (39). Note, however, that this is just another form of temporal discrimination. Again, because the time course of delivery of the capacitive-like charge is unknown, this form of temporal discrimination is also problematic.

The third problem associated with the use of large amplitude methods concerns the drastic changes in the chemical, electronic and morphological properties which occur when the film is converted from one state to the other. The reduced form of polypyrrole is a neutral, insulating, hydrophobic organic polymer. In contrast, the oxidized form is a conductive polyelectrolyte, which contains a high cationic charge density. These are very different materials and it would be surprising if they showed equivalent ion-transport rates. Because the large amplitude method converts the reduced polymer to the oxidized form (or vice-a-versa) this experiment is, in essence, conducted on one material at short times and a very different material at long times.

Finally, an examination of the chronocoulometric data for polypyrrole presented in the literature clearly shows that these data do not conform to simple theory (25). Note, for example, that Diaz et al. show Q vs. $t^{1/2}$ plots with negative intercepts; furthermore, the slopes of these plots increased as the positive limit of the potential step increased. We have obtained identical data in our laboratory (40).

The Advantages of the Small Amplitude Current-Pulse Method

The key features of the method developed here are 1.) The experiment is initiated at a potential where the polymer is in its reduced, nonconductive form (-0.4 V vs. SCE), and 2.) The total potential perturbation during the experiment is so small ($<$ 20 mV) that the polymer remains in the reduced form throughout the experiment. These features solve all of the above problems.

First, because the polymer is in the reduced form, the electrode area is just the substrate Pt area and, because this Pt electrode is a disk, diffusion must be linear. Second, because the polymer is never converted to a conductor, the large, unknown "capacitive-like" current associated with the conductive form is not observed. The only capacitive current is that associated with charging the double layer at the Pt/film interface and this can be easily measured and corrected (30). Finally, because this is a small amplitude method, the problem of converting one material into a second, very different, material during the course of the experiment is clearly obviated.

Fundamental Assumptions

This method requires three fundamental assumptions; the first might be termed the "effective medium theory assumption." Because polypyrrole is microporous (15), it is a biphasic system; (i.e., it is composed of solvent-swollen polymer and solvent-filled pores). We assume, however, that as an ion diffuses through the polymer film, it makes so many transitions between these two microphases that its motion can be described by an apparent diffusion coefficient, D_{app}. This D_{app} is the diffusion coefficient for the "effective medium" obtained by intimately mixing the two component phases. Recent work by Chien et al. clearly shows that this effective medium theory approach is valid for polypyrrole (41).

As noted above, the polymer film is an electronic insulator throughout the course of this current-pulse experiment. The question then becomes - if the polymer is not electronically conductive, how is charge transported across the film? We believe that by forcing the polymer to remain nonconductive, we convert the electronically conductive polymer into a redox polymer (42,43). In analogy to the redox polymer situation, we assume that charge (i.e, oxidized monomer equivalents) is transported via electron self-exchange with reduced polymer sites (22-24,42,43). Thus, the theories developed for diffusional electron hopping in redox polymers apply here (19,42).

Finally, we assume that the rate of this diffusional charge-transport in polypyrrole is controlled by counterion motion rather than by the actual self-exchange event. This assumption has been experimentally demonstrated for other redox polymers (44) and, as noted above, is supported by the available data on polypyrrole (14-18,45). Note, however, that even if this assumption proves to be incorrect, the D_{app} values obtained here are still valid quantitative measures of the charge-transport rate.

Background to the Current-Pulse Method

Consider a Pt working electrode which is coated with a thin, electronically conductive polymer film and immersed (along with suitable reference and counter electrodes) into a supporting electrolyte solution. At time (t) = 0 a galvanostat is used to inject a small amplitude anodic current pulse of duration τ and amplitude i_p into this electrochemical cell. At the termination of this current pulse, the working electrode is returned to open circuit and its potential, E_{oc}, is monitored as a function of time.

The anodic current pulse drives the reaction shown in Equation 1, thus increasing the concentration of Py^+ sites at the Pt/film interface. (Note, however, that the magnitude of the pulse is so small that at no time is any portion of the film converted to an electronic conductor.) The Nernst equation says that this increase in $[Py^+]$ will cause a corresponding increase in E_{oc}. After termination of the current pulse, these Py^+ sites will diffuse (i.e. self-exchange) into the bulk of the film. E_{oc} will, therefore, decrease with time; this transient will persist until the concentration of Py^+ is uniform throughout the film. Because the rate of equilibration is controlled by diffusional transport of the Py^+ sites, D_{app} can be obtained from the E_{oc} vs. t transient.

A current-pulse method was employed by Worrell et al. to obtain diffusion coefficients in intercalation compounds (46-49). These authors were able to make a number of simplifying assumptions, including 1) That semi-infinite diffusion obtained, 2.) That the diffusion layer obtained

upon termination of the current pulse was infinitesimally thin, and 3.)
That a negligible quantity of charge was introduced during the current-
pulse (46-49). If these conditions are met, the open circuit potential
after termination of the current-pulse varies linearly with the square
root of time; D_{app} can be obtained from the slope (46-49).

A Rigorous Theoretical Analysis

Because it is difficult to satisfy the above simplifying assumptions,
we were not able to apply the method of Worrell et al. to polypyrrole
films (30). A more rigorous theoretical analysis is needed if a current-
pulse method is to be applied to electronically conductive polymers.
Equations applicable to a finite diffusion situation, starting from a
diffusion layer of finite thickness, can be obtained from the heat
transfer literature (50). These equations form the basis of this new
theoretical analysis.

The details of this theoretical analysis are presented in reference
(30); only the salient features will be reviewed here. Because this
analysis is mathematically more complex, a simple relationship between E_{oc}
and time does not exist. D_{app} values are obtained by matching
experimental and simulated E_{oc} vs. t transients; D_{app} is the only
adjustable parameter (30).

The heart of this new theoretical analysis is the generation of
simulated $C_{diff,x=0}$ vs. t transients, where $C_{diff,x=0}$ is the concentration
of diffusing specie (i.e. $[Py^+]$) at the Pt/film interface (30,50,53). Two
discrete calculations are involved. First, the initial concentration-
distance profile for the diffusing species (given the symbol $C_{diff(x',\tau)}$)
must be calculated; $C_{diff(x',\tau)}$ is the diffusion layer profile extant after
termination of the current pulse, at $t=\tau$ (52,30). This initial
distribution is then plugged into the expression for the $C_{diff,x=0}$ vs. t
transient for the finite diffusion case; this equation is used to
calculate the simulated transients.

As noted above, D_{app} values were obtained by matching experimental
and simulated E_{oc} vs. t transients (30). Note, however, that the
theoretical analysis gives $C_{diff,x=0}$ vs. time transients; thus, some means
of translating from $C_{diff,x=0}$ to E_{oc} is needed. A coulometric titration
procedure was used to accomplish this translation (see Experimental
Section).

EXPERIMENTAL SECTION

Materials and Equipment

Tetraethylammonium tetrafluoroborate (99%, Aldrich) was
recrystallized from methanol and dried in vacuo at 100° C for ca. 24 hrs.
prior to use. Pyrrole (99%, Aldrich) was distilled under N_2 immediately
prior to use. Acetonitrile (UV grade, Burdick & Jackson) was used as
received. Pt foil (0.254 mm thick, Alfa) and Kel-F rod (dia = 1.27 cm,
Afton Plastics) were used to construct the Pt disk working electrodes.

All Electrochemical measurements were accomplished using an EG&G PAR
Model 173 potentiostat/galvanostat in conjunction with a PAR 175
programmer. Transients were recorded with a Nicolet 2090 digital
oscilloscope. Cyclic voltammograms were recorded on a Houston Instruments
Model 2000 X-Y recorder. Simulated data were calculated using a Compaq
Portable II computer.

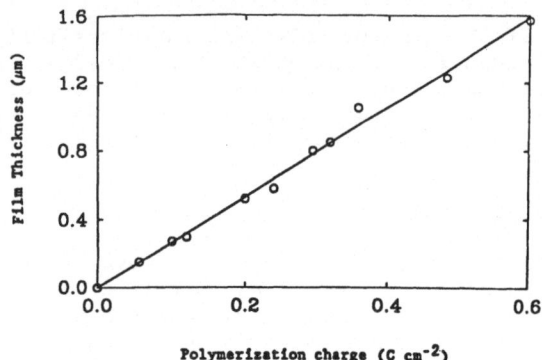

1. Plot of film thickness vs. charged consumed during polymerization for polypyrrole films.

Electrodes and Electrochemical Cells

Working electrodes were prepared as follows: Pt disks (dia = 3.3 mm) were heat-sealed onto the surface of Kel-F cylinders by melting the Kel-F around the disk with a heat gun and then physically pressing the Pt into the heat-softened surface. Electrical contact was made to the back of the Pt disk with silver epoxy (Epo-Tek 410E, Epoxy Technology). The electrodes were then polished as described previously (21).

Conventional one-compartment glass cells were employed for all electrochemical measurements. A large area, Pt gauze counter electrode (25 x 25 mm, AESAR) and a conventional saturated calomel reference electrode (SCE) were used. Potentials are reported vs. SCE. All solutions were prepared using acetonitrile as the solvent and were purged with purified N_2 prior to use.

Polypyrrole Film Deposition

Polymerization solutions were 0.5 M in pyrrole and 0.2 M in Et_4NBF_4. Pyrrole was polymerized (and deposited) galvanostatically; a current density of 1.0 mA cm^{-2} was employed. Reproducible steady state potentials of 0.84 V $\pm$ 0.01 V were observed during film deposition at this current density. After deposition, the polypyrrole film-coated electrode was transferred to monomer-free, 0.2 M Et_4NBF_4. All subsequent electrochemical measurements were performed in this electrolyte.

A Tencor Alpha Step profilometer was used to construct a calibration curve of film thickness vs. charge consumed during the polymerization process. As shown in Figure 1, film thickness is linearly related to polymerization charge. While Diaz et al. noted a similar relationship, the slope of their line (4.16 $\mu m\text{-}cm^2$ C^{-1}) was substantially greater than the slope in Figure 2 (2.64 $\mu m\text{-}cm^2$ C^{-1}), even though these films were synthesized under identical conditions (54). Film thicknesses for these studies were calculated from the known polymerization charge and the slope of the least squares fit (solid line) to the data in Figure 1.

Determination of the Relationship Between E_{oc} and $[Py^+]$

As noted in the Theory Section, this relationship allows the simulated $C_{Py^+, x=0}$ vs. t transients to be translated into E_{oc} vs. t transients. Because the current pulse experiment used to evaluate D_{app} causes the electrode potential to change from -0.4 V to at very most -0.35

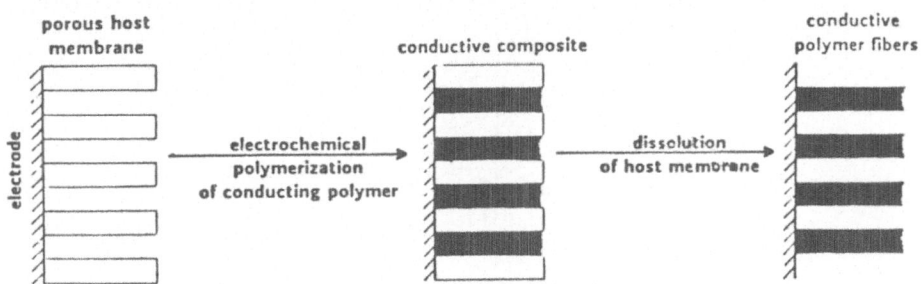

2. Schematic representation of procedure used to synthesize
fibrillar/microporous electronically conducting polymer films.

V, the relationship between E_{oc} and $[Py^+]$ needs to be defined only over
this narrow potential region. While the Nernst equation should, in
principle, define this relationship, these films are non-Nernstian (see
Figure 6). Therefore, the relationship between E_{oc} and $[Py^+]$ was
determined empirically using a coulometric titration procedure (58-60).
This procedure is described in detail in reference (30).

Determination of D_{app}

Experimental E_{oc} vs. Time Transients. Freshly synthesized polypyrrole
film was quantitatively reduced at -0.8 V vs. SCE. The reduced film was
then potentiostated at an initial potential of -0.4 V. The working
electrode was switched to open circuit and a 50 msec anodic current pulse
of the desired amplitude (100 to 400 μA cm^{-2}) was applied. The ensuing
E_{oc} vs. t transient was recorded using the Nicolet oscilloscope. These
experimental transients were then corrected by accounting for the quantity
of charge used to charge the double layer (30).

Simulated E_{oc} vs. Time Transients. Programs for generating simulated
$C_{Py+, x=0}$ vs. t transients were written in PASCAL (Turbo Pascal, Borland)
and executed on the Compaq Portable II computer. These curves were
converted to simulated E_{oc} vs. t transients. Simulated and (corrected)
experimental E_{oc} vs. t transients were plotted using LOTUS 1-2-3 (Lotus
Development Corporation) graphics. D_{app} values were obtained by manually
matching the experimental and simulated curves.

Controlling the Morphology of Electronically Conductive Polymers. Convex
Kel-F-insulated Pt disk electrodes (A = 0.5 cm^2) were constructed as
described previously (32). The convex electrode surface promoted adhesion
between the microporous membrane and the electrode surface (32). These
electrodes were pretreated as described previously (32). Nuclepore
polycarbonate microporous filtration membranes were used as the template
material (32). Membranes with pore diameters of 0.02, 1.0 and 3.0 μm were
used.

The procedure for preparing fibrillar/microporous polypyrrole
membranes is illustrated schematically in Figure 2. The membrane-coated
convex Pt disk working electrode (Figure 3) is immersed into a solution
containing the monomer (pyrrole), which is electropolymerized as described
above. Ideally (see below) polypyrrole is only synthesized in the pores
of the host membrane; a Nuclepore/polypyrrole conductive composite
membrane is obtained (Figure 2). Polymerization was terminated before the
conductive polymer fibrils reached the Nuclepore/solution interface.

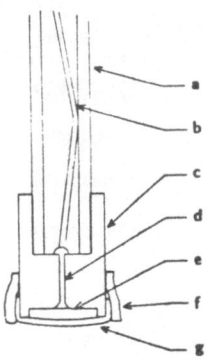

3. Schematic representation of microporous membrane-coated electrode used to synthesize fibrillar/microporous electronically conductive polymer films. a. 7mm glass tube. b. Cu wire. c. Kel-F body. d. Ag/epoxy contact. e. Convex Pt disk electrode. f. Rubber collar. g. Nuclepore microporous filtration membrane.

The conductive composite polymer membrane was then immersed in CH_2Cl_2 to dissolve away the Nuclepore (32). Two 60 min. extractions in 100 mL volumes of CH_2Cl_2 were performed. This procedure left an ensemble of conductive polymer fibrils. An electron micrograph (32) of such fibrils is shown in Figure 4.

RESULTS AND DISCUSSION

Cyclic Voltammetry of Polypyrrole

Prior to performing other electrochemical measurements, cyclic voltammetry was routinely used to assess the quality of freshly synthesized films. A typical voltammogram for a 0.27 μm polypyrrole film in 0.2 M Et_4NBF_4, MeCN is shown in Figure 5. Cyclic voltammograms for films up to ca. 1.0 μm in thickness were qualitatively similar.

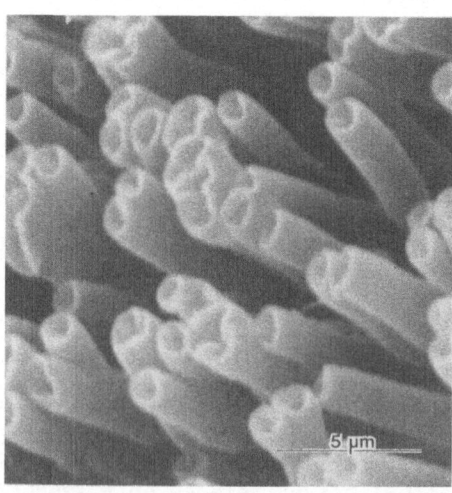

4. Scanning electron micrograph of a typical fibrillar/microporous polypyrrole film prepared using the procedure outlined in Figure 2.

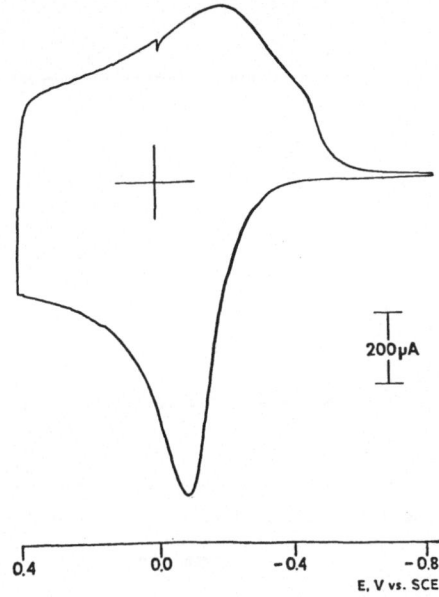

5. Typical cyclic voltammogram for 0.27 μm thick polypyrrole film. Scan rate = 10 mV sec^{-1}. Solution was 0.2 M Et$_4$NBF$_4$ in acetonitrile.

E_{oc} vs. log[Py$^+$] Calibration Curves

As discussed above, the relationship between E_{oc} and [Py$^+$] was determined empirically using a coulometric titration experiment (30). Figure 6 shows a typical semi-log (or Nernst) plot of these data. Data were obtained only over the narrow potential window accessed by the current-pulse experiment used to obtain D_{app}. Over this window E_{oc} varies linearly with log[Py$^+$]; however, the slope is super-nernstian (0.109 V/log[Py$^+$]). The data in Figure 6 were used to convert calculated $C_{Py^+, x=0}$ values into the corresponding E_{oc} values so that experimental and simulated E_{oc} vs. t transients could be compared.

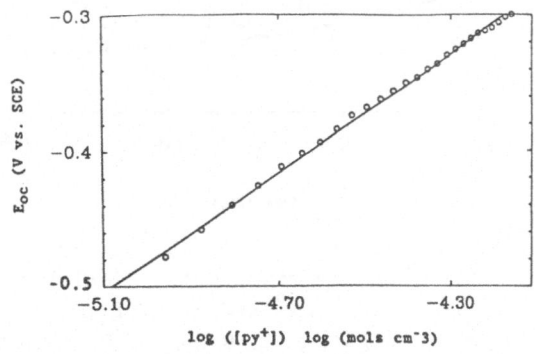

6. Typical Nernst plot for polypyrrole film-coated electrode; obtained from coulometric titration experiment. Film thickness = 0.27 μm. See text.

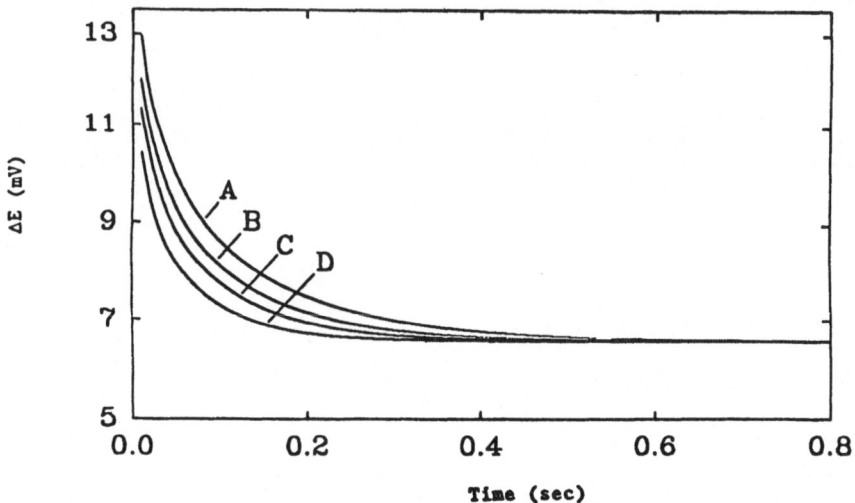

7. Simulated potential-time transients for the current-pulse experiment: Effect of magnitude of D_{app} on the transient. Simulation conditions were as follows: film thickness = 0.5 μm; t = 50 msec; i_p = 100 μA cm^{-2}; initial E = -0.4 V. D_{app}'s were as follows: A. 2×10^{-9} cm^2 sec^{-1}; B. 2.5×10^{-9} cm^2 sec^{-1}; C. 3×10^{-9} cm^2 sec^{-1}; D. 4×10^{-9} cm^2 sec^{-1}.

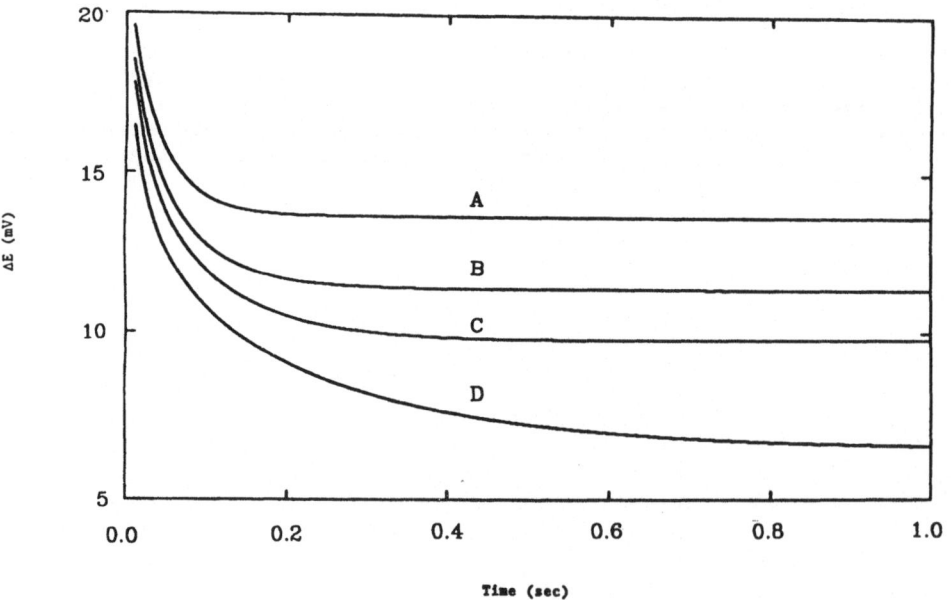

8. Simulated potential-time transients for the current-pulse experiment: Effect of film thickness on the transient. Simulation conditions as per Figure 7 except, D_{app} = 1×10^{-9} cm^2 sec^{-1} and film thicknesses as follows: A. 0.20 μm; B. 0.25 μm; C. 0.30 μm; D. 0.50 μm.

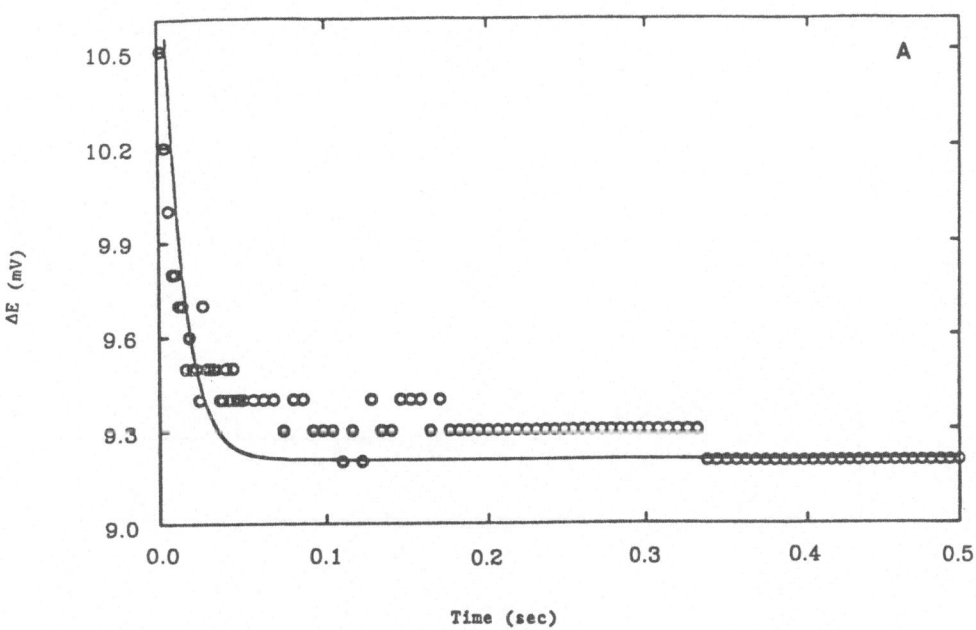

9. Comparisons of simulated (curve) and experimental (points) potential-time transients for the current-pulse experiment. Initial E = -0.4 V; i_p = 100 μA cm^{-2}; t = 50 msec; D_{app}'s are presented in Table III. Film thicknesses were as follows: A. 0.35 μm; B. 0.54 μm; C. 0.73 μm.

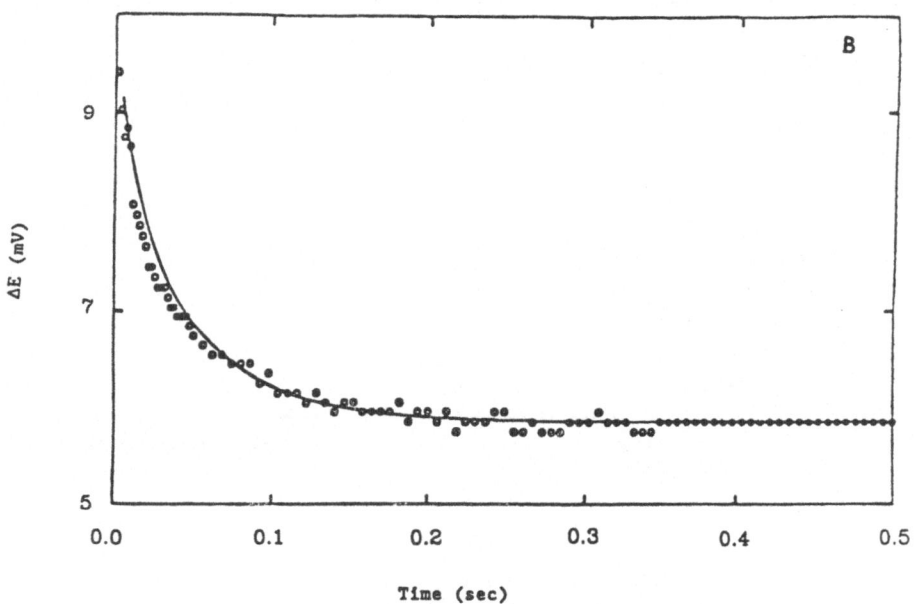

9B. (see page 129 for description)

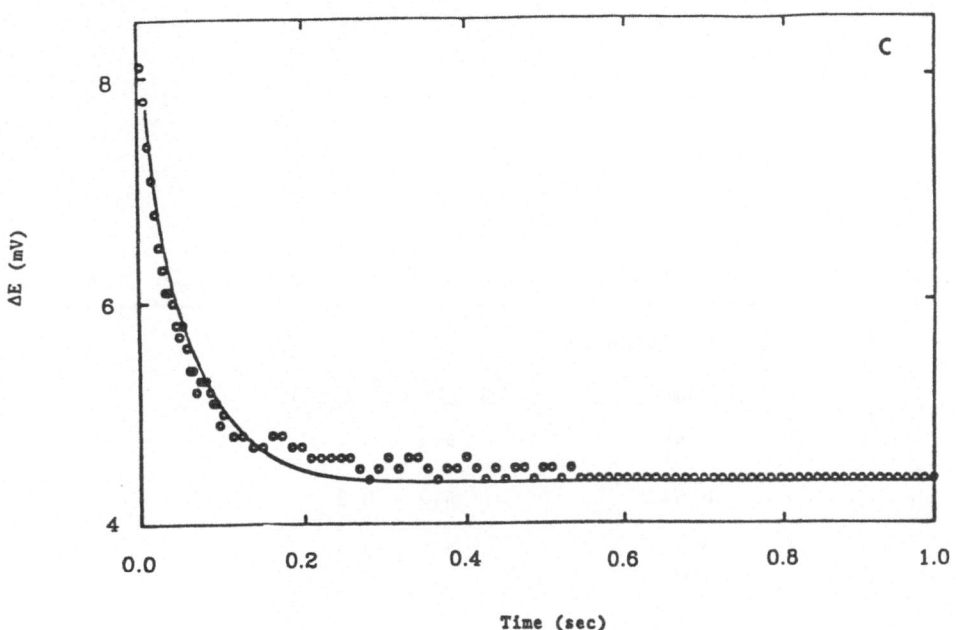

9C. (see page 129 for description)

Simulated E_{oc} vs. Time Transients

Figure 7 shows the effect of D_{app} on the simulated E_{oc} vs. t transient. The values of i_p, τ and film thickness used in Figure 7 are typical of the experiments conducted here. These simulated curves predict that the rate of film equilibration should be very sensitive to the value of the apparent diffusion coefficient. Furthermore, these data suggest that the experimental transients should occur over an easily accessible time window.

The effect of film thickness on the simulated E_{oc} vs. t transient is illustrated in Figure 8. As might be expected, both the rate of relaxation and the value of the final equilibrium potential are strongly dependent on film thickness. While this sensitivity is good in the sense

Table I. Apparent Diffusion Coefficients for Polypyrrole Films

Film Thickness μm	D_{app} x 10^9 cm^2 sec^{-1}
0.35	8.3 $\pm$ 0.3
0.54	6.2 $\pm$ 0.3
0.73	4.2 $\pm$ 0.5

that it allows the effect of film thickness on D_{app} to be accurately assessed, it places a high premium on obtaining a reliable measure of the film thickness.

Apparent Diffusion Coefficients Values

Experimental E_{oc} vs. t transients were obtained for three different polypyrrole film thickness and for four different values of the current pulse. These curves were matched to simulated curves using D_{app} as the *only adjustable parameter*. Typical data are shown in Figures 9. Excellent fits between the experimental and simulated data were obtained; this gives us some confidence in the veracity of the theoretical model.

The model can be further tested by evaluating the effect of the magnitude of the current pulse on the value of D_{app} obtained. The theory predicts that D_{app} should be independent of i_p, provided, of course, that i_p is not so large that the film is converted to an electronic conductor. To assess the effect of i_p, current pulses of 100, 200, 300 and 400 μA cm^{-2} were applied to a 0.54 μm-thick film. D_{app}'s were obtained from the best match between experimental and simulated transients. An average value of 6.2 (± 0.2) x 10^{-9} cm^2 sec^{-1} was obtained. Since this standard deviation is identical to the standard deviation for replicate measurements at a single value of i_p, there is clearly no dependence of D_{app} on i_p.

D_{app} values for three different polypyrrole film thicknesses are shown in Table I. Each D_{app} represents an average of eight determinations (two films, four different current densities for each film). Note first that the D_{app} values reported in Table I are as much as an order of magnitude larger than values reported by Diaz et al. (25) and Osaka et al. (29). Because these previously reported data were obtained using large amplitude chronoamperometric methods, we believe (see Theory Section) that these D_{app} values are suspect.

The theory developed here predicts that, if thin films have the same chemical and morphological composition as thick films, D_{app} should be independent of film thickness. In contrast, while excellent matches between experimental and simulated transients were obtained for all of the film thickness reported here (Figure 9), D_{app} clearly decreases with film thickness (Table I). These data are in agreement with the often reported qualitative observation that charge transport rates decrease with film thickness (45). Furthermore, Osaka et al. observed a similar trend in their data (29).

Fibrillar/Microporous Polypyrrole Promotes Higher Rates of Charge-Transport Than Conventional Polypyrrole Films

An electron micrograph of typical polypyrrole fibrils prepared via the microporous membrane-coated electrode technique, is shown in Figure 4. We prepared such fibrillar/microporous films because we believed that the rate of charge-transport in these films would be faster than in conventional polypyrrole. We postulated that the fibrillar/microporous films would support higher rates of charge-transport because 1) The micropores, when filled with solvent, would become fast ion-conducting channels into the film, 2) While counterions would ultimately have to enter or exit the polymer phase (Equation 1), these ions would only have to traverse the radius of the narrow fibril, and 3) Transport into the polymer phase would be changed from a linear diffusion process (conventional polypyrrole film) to a cylindrical diffusion process (fibrous film).

To test whether the fibrous versions of polypyrrole supported higher rates of charge-transport than conventional polypyrrole films, charge-time transients were obtained for both types of polypyrrole. These charge-time transients were associated with the reduction of the polymer. The polymer films were first quantitatively oxidized at +0.2 V vs. SCE. The potential was then stepped to -0.6 V where the reduction of the polymer proceeds at the diffusion-controlled rate. The corresponding charge vs. time transients were recorded using the X-Y recorder.

Reductive charge-time transients were obtained for a conventional (amorphous) polypyrrole film and for fibrillar/microporous films having fibril diameters of 3, 1 and 0.2 μm. It is important to point out that

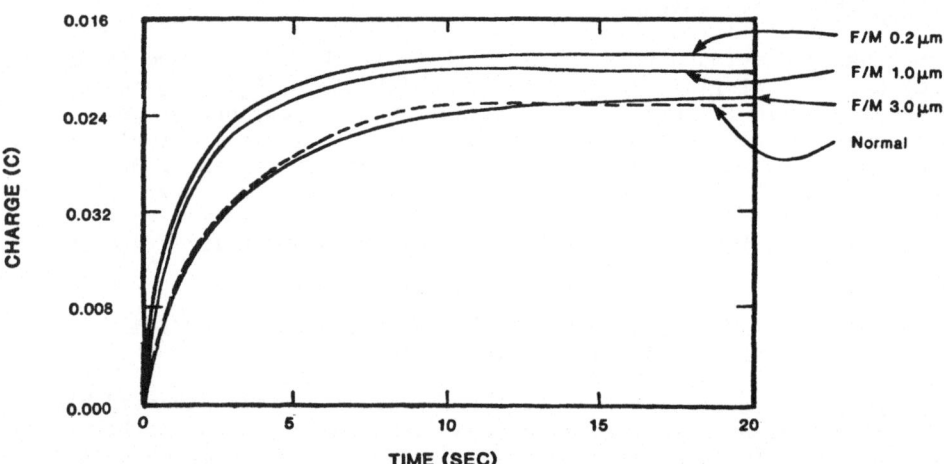

10. Charge-time transients associated with the reduction of a conventional polypyrrole film (1.6 μm thick; dashed curve labeled "Normal") and three fibrillar/microporous polypyrrole films (labeled "F/M"; fibril diameters as shown.

all four of these films contained the same quantity of polypyrrole. This was assured by using 1.2 C of charge during the synthesis of all of the films.

The charge-time transients are shown in Figure 10 ("normal" in this figure refers to the conventional polypyrrole film; this film was ca. 1.6 μm thick; the designator F/M refers to the fibrillar/microporous membranes). Note first that the charge-time transients for the conventional (normal) film and for the fibrillar/microporous film with 3.0 μm-diameter fibrils are essentially identical. This is not surprising since the radii of these fibrils are essentially identical to the thickness of the conventional film.

Note, however, that at any time, the 0.2 and 1.0 μm fibril-diameter films <u>deliver more charge</u> than the conventional films. This is very clear evidence that these narrow fibril-diameter films promote faster rates of charge-transport than conventional polypyrrole. Furthermore, the total amount of charge delivered by these fibrous versions of polypyrrole is greater than the charge delivered by the conventional polypyrrole film. This observation has important implications to possible battery applications of these polymers.

There is, however, one disappointing feature about the data in Figure 10. While the rate of charge-transport for the 1.0 μm fibril-diameter film is substantially higher than for the normal film, the 0.2 and 1.0 μm fibril-diameter films support very similar rates of charge-transport. We had predicted (and hoped) that the rate of charge-transport would become faster and faster as the fibril diameter was decreased; the data in Figure 10 suggest that this is not happening.

Careful electron microscopic investigation of the bases of the polypyrrole fibers studied here revealed that these fibers do not grow directly from the substrate Pt electrode. Rather, a very thin base layer of conventional polypyrrole forms first; the fibrils grow out of this base layer. This base layer undoubtedly results from the fact that there is always a thin layer of solution between the membrane and the substrate electrode surface. This base layer undoubtedly retards the rate of charge-transport in the fibrous versions of the polymer.

The above discussion suggests that superior fibrillar/microporous films would be obtained if the base layer of conventional polypyrrole could be eliminated. We have recently shown that this can be accomplished by sputtering the electrode directly onto the back of the host membrane (55); we will report the results of these studies in a forthcoming manuscript (55).

CONCLUSION

We have developed a new small amplitude electrochemical method for determination of apparent diffusion coefficients associated with charge-transport in electronically conductive polymers. For the reasons noted in the Theory Section, this method is ideally suited for investigations of such polymers. In a more general sense, this method should also be a convenient and powerful tool for studying transport in nearly any type of electroactive polymer film. Some of the features which make this an attractive and powerful method for analysis of transport in polymer films are discussed below.

First, because this is a small amplitude method, the electrochemical, chemical and morphological properties of the film are not greatly perturb by the analysis. Because electroactive polymer films may show dramatic changes in solvent content, ionic character, morphology, etc. upon oxidation or reduction, the minimally invasive character of this method is an attractive feature. Second, and also related to the small amplitude nature of this method, by starting at different values of initial applied potential, the effects of the extent of oxidation or reduction on charge transport in the film can be assessed. Thus, for example, the effects of redox-induced changes in solvent content or morphology can be quantitatively evaluated using this method. This is not possible with large amplitude methods.

Third, unlike conventional methods for evaluating charge transport in polymer films on electrode surfaces (19-24), there is no current flow during the data acquisition period. Thus, effects of migration and uncompensated film resistance are minimized or completely eliminated. Fourth, the theoretical model for obtaining D_{app} is applicable to both the semi-infinite and finite diffusion cases. Thus, unlike many other methods (19-24), it is not necessary to adjust the experimental conditions such that the semi-infinite case is obtained.

We have also shown that fibrillar/microporous versions of polypyrrole support faster rates of charge-transport than conventional polypyrrole films.

Acknowledgements. This work was supported by the Office of Navel Research and the Air Force Office of Scientific Research.

References

1. *Handbook of Conducting Polymers*; Skotheim, T. A., Ed.; Marcel Dekker: New York, 1986.

2. Burgemayer, P.; Murray, R. W. *J. Am. Chem. Soc.* **1982**, 104, 6139.

3. Burgemayer, P.; Murray, R. W. *J. Phys. Chem.* **1984**, 88, 2515.

4. Garnier, F.; Tourillon, G.; Gazard, M.; Dubois, J. C. *J. Electroanal. Chem.* **1983**, 148, 299.

5. Gazard, M. in *Handbook of Conducting Polymers*; Skotheim, T. A., Ed.; Marcel Dekker: New York, 1986; Vol. 1, Chapter 19.

6. Kaufman, J. H.; Chung, T.-C.; Heeger, A. J.; Wudl F. *J. Electrochem. Soc.* **1984**, 131, 2092.

7. Mermilliod, N.; Tanguy, J.; Petiot, F. *J. Electrochem. Soc.* **1986**, 133, 1073.

8. Button, P.; Mastragostino, J.; Panero, S.; Scrosati, G. *Electrochimica ACTA* **1986**, 31, 783.

9. White, H. S.; Kittlesen, G. P.; Wrighton, M. S. *J. Phys. Chem.* **1985**, 89, 5133.

10. Kittlesen, G. P.; White, H. S.; Wrighton, M. S. *J. Am. Chem. Soc.* **1984**, 106, 7389.

11. Thackeray, J. W.; White, H. S.; Wrighton, M. S. *J. Phys. Chem.* **1985**, 89, 5133.

12. Thackeray, J. W.; Wrighton, M. S. *J. Phys. Chem.* **1986**, 90, 6674.

13. Chao, S.; Wrighton, M. S. *J. Am. Chem. Soc.* **1987**, 109, 2197.

14. Kanazawa, K. K.; Diaz, A. F.; Geiss, R. F.; Gill, W. D.; Kwak, J. F.; Logan, J. A.; Rabolt, J. F.; Street, G. B. *J. Chem. Soc., Chem. Commun.* **1979**, 854.

15. Bull, R. A.; Fan, F.-R.; Bard, A. J. *J. Electrochem. Soc.* **1982**, 129, 1009.

16. Kaufman, J. H.; Kanazawa, K. K.; Street, G. B. *Phys. Rev. Lett.* **1984**, 53, 2461.

17. Miller, L. L.; Zinger, B.; Zhou, X.-Q. *J. Am. Chem. Soc.* **109**, 2267 (1987).

18. Buttry, D. A. Presented at the 193rd ACS Meeting, Denver, CO, April, 1987.

19. Peerce, P. J.; Bard, A. J. *J. Electroanal. Chem.* **1980**, 114, 89.

20. Martin, C. R.; Rubenstein, I.; Bard, A. J. *J. Am. Chem. Soc.* **1982**, 104, 4817.

21. Martin, C. R.; Dollard, K. A. *J. Electroanal. Chem.* **1983**, 159, 127.

22. Oyama, N.; Anson, F. C. *J. Electrochem. Soc.* **1980**, 127, 247.

23. Shigehara, K.; Oyama, N.; Anson, F. C. *J. Am. Chem. Soc.* **1981**, 103, 2552.

24. Oyama, N.; Shimomura, T.; Shigehara, K.; Anson, F. C. *J. Electroanal. Chem.* **1980**, 112, 271.

25. Genies, E. M.; Bidan, G.; Diaz, A. F. *J. Electroanal. Chem.* **1983**, 149, 101.

26. Marque, P.; Roncali, J.; Garnier, F. *J. Electroanal. Chem.* **1987**, 218, 107.

27. Nagasubramanian, G.; Di Stefano, S.; Moacanin, J. *J. Phys. Chem.* **1986**, 90, 4447.

28. Chiba, C.; Ohsaka, T.; Oyama, N. *J. Electroanal. Chem.* **1987**, 217, 239.

29. Osaka, T.; Naoi, K.; Ogana S.; Nakamura, S. *J. Electrochem. Soc.* **1987**, 134, 2096.

30. Penner, R.M.; Van Dyke, L.S.; Martin, C.R. J. Phys. Chem. 1988, 42, 5274.

31. Feldburg, S. W. *J. Am. Chem. Soc.* **1984**, 106, 4671.

32. Penner, R.M.; Martin, C.R. J. Electrochem. Soc. 1986, 133, 2206.

33. As we shall see, the discussion presented applies to chronocoulometry and other large amplitude methods also.

34. Bard, A. J.; Faulkner, L. R.; *Electrochemical Methods: Theory and Applications*; John Wiley and Sons: New York, 1980, p 143.

35. The question of available electrode area arises because in the oxidized form both the substrate Pt surface and some fraction of the internal surface area of the electrode are available for polymer electrolysis. In contrast, in the reduced form only the substrate Pt electrode is available for polymer electrolysis.

36. Penner, R. M.; Martin, C. R.; *J. Electrochem. Soc.* **1986**, 133, 310.

37. Tanguy, J.; Mermilliod, N.; Hoclet, M. *J. Electrochem. Soc.* **1987**, 134, 795.

38. Tanguy, J.; Mermilliod, N.; Hoclet, M. *Synth. Met.* **1987**, 18, 7.

39. Bard, A. J.; Faulkner, L. F. *Electrochemical Methods: Theory and Applications*; John Wiley and Sons: New York, 1980, p 200.

40. Penner, R. M. unpublished results, Texas A&M University, Dec., 1987.

41. Schlenoff, J. B.; Chien, J. C. W. *J. Am. Chem. Soc.* **1987**, 109, 6269.

42. Laviron, E. *J. Electroanal. Chem.* **1980**, 112, 1.

43. Chidsey, C. E. D.; Murray, R. W. *Science* **1986**, 231, 25.

44. Jernigan, J. C.; Murray, R. W. *J. Am. Chem. Soc.* **1987**, 109, 1738.

45. Diaz, A. F.; Castillo, J. I.; Logan, J. A.; Lee, W.-Y. *J. Electroanal. Chem.* **1981**, 129, 115.

46. Basu S.; Worrell, W. L. in *Fast Ion Transport in Solids*, Vashishta, P.; Mundy, J. N.; Shenoy, J. K., Eds, Elsevier North Holland: Amsterdam 1979.

47. Basu, S.; Worrell, W. L. in *Proceedings of the Symposium on Electrode Materials and Process for Energy Conversion and Storage*, McIntyre, J. D. E.; Srinivasan, S.; Will, F. G., Eds, The Electrochemical Society: Princeton, 1977; p 861.

48. Nagelburg, A. S.; Worrell, W. L. in *Proceedings of the Symposium on Electrode Materials and Process for Energy Conversion and Storage*, McIntyre, J. D. E.; Srinivasan, S.; Will, F. G., Eds, The Electrochemical Society: Princeton p 861 (1977).

49. Nagelburg, A. S.; Worrell, W. L. *The Electrochemical Society Extended Abstracts*, Vol 77-1, Philadelphia, PA, 1977; 948.

50. Carslaw, J. S.; Jaeger, J. C. *Conduction of Heat in Solids*; Oxford University: London, 1959.

51. The parameter x is the distance from the electrode/film interface. Thus, x extends from x=0 to x=d where d is the film thickness.

52. The symbol x' is used to express distance from the electrode/film interface for the <u>initial</u> diffusion layer profile.

53. Carslaw, J. S.; Jaeger, J. C. *Conduction of Heat in Solids*; Oxford University: London, 1959, p 263.

54. Díaz, A.F.; Castillo, J.J. J. Chem. Soc. Chem. Commun. 1980, 397.

55. Van Dyke, L.S. Unpublished results Texas A&M University, October, 1988.

FUNCTIONAL POLYMERS

Otto Vogl

Polytechnic University
333 Jay Street
Brooklyn, N.Y. 11201

Functional polymers are the basis for the most important trends in polymer science in the last decade. They have properties that are not only derived from the macromolecular structure, but depend to a significant extent, or even entirely, on the functional group substituents on the macromolecules.

Increased sophistication in the use of advanced polymeric materials requires that the structure of such polymers be carefully designed (speciality polymers). The high demand on the design and the actual tailor-making of such macromolecular materials require a great deal of imagination and detailed knowledge in synthesis and structure/property relationships (macromolecular architecture and macromolecular engineering). In the last decade, research in polymer chemistry and production in the polymer-related industries have shifted from the emphasis on polymers based on raw material availability and high cost efficiency to market- and use-oriented tailor-made polymeric materials. Earlier efforts were concerned with the most efficient and economical preparation of monomers (primarily from oil) and their subsequent polymerization; these advances were the basis for the vigorous development and the commercial success of large-scale commodity plastics (polyolefins, polystyrene, and poly(vinyl chloride), as well as condensation polymers such as polyesters, polyamides, and the early examples of engineering plastics). This vigorous progress of polymer science is the reason that our era is often referred to as the PLASTIC AGE. The change in the progress of polymer science from polymers with a high volume of production and with a relatively low economic return (requiring no significant research efforts) to the preparation and production of low-volume polymeric materials, yet with economically high return, provides new dimensions for new growth in design and utilization of macromolecules.

This emphasis on the functionality in the macromolecule rather than on the macromolecular structure itself also means that scientists working with these materials need to have new combinations of knowledge to be successful. Not only do they need to know how to design macromolecular structures, but also what structures are required to give the desired combination of polymer properties. And how can such materials be fabricated? Should it be done by traditional methods such as melt processing, or by monomer or oligomer casting techniques; or, for high-temperature-resistant polymers (materials with properties close to those of ceramic materials), should they be handled by modified metal working techniques?

Another question important for the academic community is: What type of training should we provide at universities for the polymer scientists which we need at the turn of the century? This is essential to provide the proper manpower which can effectively handle delicate design of polymer structures for advanced applications which are based on macromolecular design, macromolecular architecture, and macromolecular engineering.

In the design of macromolecular structures with functional groups it is not only necessary to be concerned with the macromolecule and the functional group, but it is becoming of further importance to be concerned with the spacing of the functional groups with respect to the macromolecular backbone chain. Nature has carefully designed natural macromolecular structures and has placed functional amino acid units with spacer groups in strategic positions in proteins and specific functional groups in sugar units in polysaccharides to obtain macromolecular structures with optimal biological activity. With clever structure design, sequence, and spacer arrangements, nature has designed enzymes, biologically and immunologically active macromolecular structures. Much could be done in the design of synthetic maromolecular with the proper knowledge of the intricacies and interrelations of macromolecular backbone chains, functionalities, and spacer groups.

The accurate historic origin of the expression "functional polymers," as in most cases in the naming of a scientific discipline, cannot be reconstructed unequivocally. Takemoto used "functional monomers and polymers" as the general theme for the publication of his work on polymerizable derivatives of nucleic bases, purines, and pyrimidines. Probably simultaneously, the expression "functional polymers" was used by our research group as a common expression for polymeric stabilizers and polymeric drugs.

At the meeting of the steering committee (which included Vogl, Furukawa, and Saegusa) in Hakone, Japan, 1974, during the first U.S.-Japan Seminar, the title "Functional Polymers" was adopted as the theme for the second U.S.-Japan Seminar in Pingree Park, Colorado, 1978. This consensus was reached after a lengthy discussion, and it was concluded that the field of functional polymers was the most promising direction in polymer research and consequently the best subject for the next U.S.-Japan Seminar. This decision became the nucleus for the most exciting thrust in polymer research in the following decade; the results of these investigations provided the scientists of the first Seminar and their colleagues with ample opportunities for presentations at the second Seminar. History has proved that the selection of Functional Polymers as the theme for the second Seminar was a timely decision, and the ensuing years saw vigorous research on functional polymers and a new direction in polymer science.

Research on functional polymers in the U.S. showed steady progress; in Japan, research increased rapidly, and within one year, 10-12 research groups were working on functional polymers. Today, there are two Departments of Functional Polymers in Japanese universities. Interest in functional polymers is also reflected in the selection of this subject for keynote and plenary lectures at highly prestigious conferences on polymer science. In two of the last three IUPAC Symposia on Macromolecular Chemistry, the plenary lectures were on functional polymers; symposia and major conferences on functional polymers are now commonplace and worldwide. There was a major conference in 1981 on functional polymers in Kunming, P.R.C., and the symposia Macromolecules '86 and '89 in Oxford; two symposia on specialty polymers in the U.K. and the U.S. were primarily concerned with functional polymers. Other conferences with functional polymers as the focal point stressed the ideas that were first formulated almost two decades ago, ideas that have helped to establish this now well recognized field of polymer science.

SOLUBLE POLYMER-BOUND CATALYSTS

David E. Bergbreiter

Department of Chemistry

Texas A&M University

College Station, Texas 77843

Introduction

Much of the impetus for using polymers as reagents and catalysts derives from the success of Merrifield's procedure for peptide synthesis. The insolubility of divinylbenzene-crosslinked polystyrenes, their facile chemical derivatization and desirable physical properties all combined to make these polymer-supported reagents the reagents of choice for synthesis of small peptides.[1] In the late 1960's and throughout the 1970's there was a great deal of interest in extending this chemistry to include ordinary organic reactions. While the objectives in each individual case varied, they generally included one or more of the following goals:

1. Recovery of a valuable reagent
2. Separation of a toxic or noxious reagent
3. Enhanced stability for a reagent or intermediate
4. Quantitative separation of reagents from reaction products
5. Kinetic isolation of a reagent or intermediate

Achieving any of these goals depends in a critical way on the properties of the polymer support. In essentially all cases, crosslinked polystyrenes were used as the polymer support. This particular support was readily available either as Merrifield resin, ion exchange resins and in the form of underivatized resin. Crosslinked polystyrene was also useful because it was an insoluble polymer matrix which was compatible with the organic solvents normally used in synthesis and homogeneous catalysis. Indeed, through the initial efforts of a number of groups it was shortly shown that crosslinked polystyrene's insolubility and stability coupled with the diffusional restrictions associated with insoluble macromolecules allowed for at least partial success in achieving each of the above goals.

By the 1980's, insoluble polymer-bound reagents and catalysts had achieved wide recognition and were, in many cases, commercially available.[2-4] However, while insoluble polymers like these were successful as reagent and catalyst supports, it also became apparent that there are limitations associated with such species. In particular, the insolubility of the polymer both *during* and at *the end* of the reaction can be disadvantageous. Such insolubility makes formation of the reagent or catalyst more tedious and makes reagent or catalyst characterization much more difficult. Second, while there are instances in which diffusional restrictions are desirable and necessary, such restrictions often limited the range of substrates or a polymer-bound reagents utility. Finally, while insoluble polymer-bound reagents and catalysts were modeled after homogeneous reagents and catalysts, they often behaved somewhat differently. Sometimes they had different selectivity. Generally they had lower reactivity. Some of the most interesting results with these catalysts came from cases where "site isolation" led to improved catalyst stability. Regardless, the original goal of making an insoluble polymer-bound catalyst which mimiced its homogeneous counterpart proved to be an elusive goal.

An alternative to insoluble polymer-bound reagents or catalysts which was explored by Bäyer's group beginning in the 1970's and then thereafter by other groups was the use of soluble polyethylene glycol or polystyrene derivatives as a ligands, reagents or catalyst supports.[5,6] As was originally true for polystyrene-based reagents, Bäyer's group first studied peptide synthesis. They developed a methodology for peptide synthesis which they termed a "liquid phase peptide synthesis" strategy using poly(ethylene glycol) as the polymer which supported the growing peptide chain. The general approach was fundamentally similar to that of Merrifield. The difference was that the reactions were carried out homogeneously. Characterization of the intermediate products was also facilitated by the polymer-peptide's solubility. While this approach was successful, it too had problems. In particular, it suffered from the fact that the eventual product of the repetitive peptide synthesis was essentially a diblock polymer. In the original polymer, the poly(ethylene glycol) chain was of sufficient size to determine the polymer's physicochemical properties which were important in this synthesis. However, as the peptide synthesis proceeded and the peptide portion of the chain increased in size, it could begin to affect important properties like solubility.

The use of soluble polymers as catalysts was also explored by Bäyer.[6] His group showed that both diphenylphosphinated polystyrene and diphenylphosphinated poly(ethylene glycol) could be used as recoverable, reusable hydroformylation catalysts. Separation of the catalyst and the reaction products in these cases was achieved by taking advantage of the properties of the polymer chain. Solvent precipitation or membrane filtration both proved to be acceptable techniques to isolate products free from the polymer-bound catalyst.

Our interest in using the properties of linear soluble polymers to make more useful catalysts or reagents developed from a somewhat different direction. Although we had known of Bäyer's original work and of some of the other sporadic reports of soluble polymers being used in synthesis or catalysis, we became especially interested in this area because of other work in our group on polyethylene derivatization. In this project on polyethylene chemistry, we had developed a new procedure for polyethylene derivatization we termed "entrapment functionalization".[7,8] In this approach, we first prepared a terminally functionalized ethylene oligomer or a terminally functionalized ethylene-butene cooligomer. Polyethylene derivatization was achieved by dissolving both the oligomer of interest and high density polyethylene in toluene or xylene at 100 °C and then allowing the mixture of macromolecules to coprecipitate. While this particular chemistry was not designed to be useful in catalysis (the effective molar concentration of oligomer was $< 10^{-4}$ M), the fact that oligomers could be recovered from solution suggested applications of this chemistry in catalysis. Further preliminary work using ESR spectroscopy showed that spin labeled linear oligomers like **1** with more than 150 carbons were quantitatively recovered from solution even in the absence of added host polymer lending further support to the idea that similar ligands could be useful in catalysis. The application of functional polymers similar to **1** as recoverable, recyclable ligands in homogeneous transition metal catalyzed reactions is the subject of this chapter.

Ligand Syntheses

Our first efforts to exploit this polyethylene chemistry in a catalytic reaction were designed so that we could test the utility of homogeneous polyethylene-bound catalysts in comparison to known conventionally-ligated homogeneous catalysts. We were especially interested in showing that the homogeneous catalysts containing a soluble polyethylene-based ligand had activity and selectivity which could be varied

1

and controlled by ligands in a manner analogous to that of homogeneous catalysts with normal ligands. This required that we develop methods to prepare polyethylene oligomers terminated with groups that could easily bind transition metals. Our initial efforts therefore focused on the synthesis of a variety of phosphite and phosphine terminated polyethylene oligomers which we could use as ligands.

The synthesis of terminally functionalized ethylene oligomers with the desired solubility properties proved to be relatively simple using anionic oligomerization of ethylene. Anionic oligomerization of butadiene had been used previously to prepare polybutadiene oligomers which can be used to prepare polyethylene oligomers.[9-11] However, these polybutadiene oligomers had to be hydrogenated to form polyethylene oligomers and the product saturated oligomers also contained some ethyl branching due to the unavoidable presence of some 1,2-polybutadiene. The alternative of anionic oligomerization of ethylene proved to be more useful (equation 1).[12] Anionic polymerization of ethylene is known. However, it is of comparatively little interest because it is not a feasible route to high molecular weight polymers. The living oligomer **2** precipitates during the polymerization when the reaction is carried out at room temperature and the resulting heterogeneous oligomerization is too inefficient to prepare high molecular weight materials. However, the modest weight oligomers produced (M_N ranges from 1200 - 4500) are quite adequate for our purposes. In addition, anionic polymerization of ethylene is experimentally simple and affords a readily derivatized living oligomer. Unlike anionic polymerizations designed to prepare high molecular weight materials, these oligomerizations can be carried out with conventional inert atmosphere techniques since some inadvertent termination due to adventitious water or oxygen can be tolerated. Any by-product proton terminated oligomer will either not affect the eventual catalysis or will be lost to solution during an eventual precipitation step.

$$CH_2{=}CH_2 \xrightarrow[\text{(CH}_3)_2\text{NCH}_2\text{CH}_2\text{N(CH}_3)_2]{\text{BuLi}} Bu{-}CH_2CH_2{+}CH_2CH_2{+}_n CH_2CH_2Li \qquad (1)$$

2

Derivatization of the living oligomer **2** was carried out in several ways all of which relied on the chemical similarity of **2** and an alkyllithium reagent. In some cases (eqs. 2 and 3), electrophilic substitution of **2** led directly to ligands. In other cases, further synthetic manipulations were required. These reactions are summarized in the discussion below.

While some reactions above directly produced the desired ligand or catalyst, further derivatization of the initial oligomer was required in other cases. For example, the initially formed hydroxy terminated oligomer **5** derived either directly from ethylene oxide quench of **2** or from reduction of the carboxylic acid **3** (equations 4 and 5) was transformed into an polyethylene diphenylphosphine **4** ligand by initial conversion to a methane sulfonate ester **6** followed by nucleophilic substitution using lithium diphenylphosphide (equation 6). The oligomer **5** was also converted into a phosphite group **7**. The best procedure for this transformation involved deprotonation with *n*-butyllithium to form a lithium alkoxide followed by exchange with excess

$$2 \quad \xrightarrow[\text{2. H}_3\text{O}^+]{\text{1. CO}_2} \quad \text{Bu}-\text{CH}_2\text{CH}_2\text{(}\text{CH}_2\text{CH}_2\text{)}_n\text{CH}_2\text{CH}_2\text{CO}_2\text{H} \qquad (2)$$

$$\mathbf{3}$$

$$2 \quad \xrightarrow{\text{1. ClP(C}_6\text{H}_5)_2} \quad \text{Bu}-\text{CH}_2\text{CH}_2\text{(}\text{CH}_2\text{CH}_2\text{)}_n\text{CH}_2\text{CH}_2\text{P(C}_6\text{H}_5)_2 \qquad (3)$$

$$\mathbf{4}$$

$$2 \quad \xrightarrow[\text{2. H}_3\text{O}^+]{\text{1. }\triangle\text{O}} \quad \text{Bu}-\text{CH}_2\text{CH}_2\text{(}\text{CH}_2\text{CH}_2\text{)}_n\ \text{CH}_2\text{CH}_2\text{OH} \qquad (4)$$

$$\mathbf{5}$$

$$3 \quad \xrightarrow[\text{2. CH}_3\text{OH}]{\text{1. BH}_3\cdot\text{S(CH}_3)_2} \quad \text{Bu}-\text{CH}_2\text{CH}_2\text{(}\text{CH}_2\text{CH}_2\text{)}_n\text{CH}_2\text{CH}_2\text{OH} \qquad (5)$$

$$\mathbf{5}$$

triarylphosphite (equation 7). Alternative procedures which involved reaction of the oligomeric alcohol **5** with ClP(OR)$_2$ PCl$_3$ proved more troublesome in that significant amounts of phosphonate often formed. A minor disadvantage of these procedures is that they all produce the electronic equivalent of an alkyldiaryl phosphine or phosphite. While these ligands are actually satisfactory, the electronic equivalent of triarylphosphine or -phosphite ligands were also accessible by variations of these procedures (equations 8 and 9). While alkyllithium reagents are often poor carbon nucleophiles for carbon-carbon bond formation with alkyl halides, these reagents react readily with benzyl halides to form carbon bonds to the benzyl carbon. This reaction, which has been used as a titration procedure for analysis of residual base in alkyllithium reagent chemistry, was successfully applied to oligomers like **2**. By using benzyl bromides which possessed a second halide or protected phenolic group, both triarylphosphines (**8**) and triarylphosphites (**9**) were accessible.

$$5 \quad \xrightarrow[\text{2. LiP(C}_6\text{H}_5)_2]{\substack{\text{1. CH}_3\text{SO}_2\text{Cl, Et}_3\text{N}\\ \text{CH}_2\text{Cl}_2}} \quad \text{Bu}-\text{CH}_2\text{CH}_2\text{(}\text{CH}_2\text{CH}_2\text{)}_n\ \text{CH}_2\text{CH}_2\text{PPh}_2 \qquad (6)$$

$$\mathbf{4}$$

$$5 \quad \xrightarrow[\text{2. (ArO)}_3\text{P}]{\text{1. BuLi}} \quad \text{Bu}-\text{CH}_2\text{CH}_2\text{(}\text{CH}_2\text{CH}_2\text{)}_n\text{CH}_2\text{CH}_2\text{OP(OAr)}_2 \qquad (7)$$

$$\mathbf{7}$$

A particular advantage of the syntheses using soluble polyethylene oligomers over syntheses using insoluble polystyrene is the facility with the products of the polyethylene chemistry can be characterized. While some NMR data can often be obtained from NMR spectra of suspensions of polystyrene-bound ligands, the ligands and ligand precursors **3-9** were all readily characterized by high resolution state ^{1}H NMR spectroscopy in toluene-d_8 at 100 °C. Thus it is comparatively easy to assay the course of these synthetic transformations and to verify the chemical integrity of the ligand both before and after a reaction.

$$2 \xrightarrow[\text{2. LiP(C}_6\text{H}_5)_2]{\text{1. BrCH}_2\text{C}_6\text{H}_4\text{Br}} \text{Bu} - \text{CH}_2\text{CH}_2 \left(\text{CH}_2\text{CH}_2 \right)_n \text{CH}_2\text{CH}_2 \text{—} \underset{\mathbf{8}}{\text{C}_6\text{H}_4} \text{—} P(\text{C}_6\text{H}_5)_2 \qquad (8)$$

$$2 \xrightarrow[\substack{\text{2. (CH}_3)_3\text{SiI} \\ \text{3. BuLi} \\ \text{4. (ArO)}_3\text{P}}]{\text{1. BrCH}_2\text{C}_6\text{H}_4\text{OCH}_3} \text{Bu} - \text{CH}_2\text{CH}_2 \left(\text{CH}_2\text{CH}_2 \right)_n \text{CH}_2\text{CH}_2 \text{—} \underset{\mathbf{8}}{\text{C}_6\text{H}_4} \text{—} P(\text{OAr})_2 \qquad (9)$$

Catalyst Synthesis

Catalyst synthesis was generally accomplished by procedures analogous to those used to make the conventionally ligated catalyst. For example, addition of diphenylphosphinated polyethylene to the a solution of the [chlororhodium(ethylene)] dimer produced an analog of Wilkinson's complex.[13] However, it was also possible to directly exchange the polyethylene diphenylphosphine for triphenylphosphine. This exchange procedure is illustrated by the catalysts syntheses shown below. In both the rhodium(I) and palladium(0) examples shown, the exchange reaction is driven to completion by precipitation of the oligomerically ligated catalyst. While we have not been able to exclude the possibility that some triphenylphosphine is also present as a ligand, the low leach rates seen suggest that little or no triphenylphosphine remains as a ligand. This may reflect the better σ-basicity of the polyethylene diphenylphosphine ligand. Alternatively, there could be some cooperative effect operating which favors polymer-bound complex formation.

$$\underset{\mathbf{4}}{\text{PE} - \text{PPh}_2} + [(\text{C}_2\text{H}_4)\text{RhCl}]_2 \longrightarrow \underset{\mathbf{10}}{[\text{PE} - \text{PPh}_2]_3\text{RhCl}} \qquad (10)$$

$$\mathbf{4} + \underset{\mathbf{11}}{[(\text{C}_6\text{H}_5)_3\text{P}]_3\text{RhCl}} \longrightarrow \mathbf{10} + (\text{C}_6\text{H}_5)_3\text{P} \qquad (11)$$

$$\mathbf{4} + \underset{\mathbf{12}}{[(\text{C}_6\text{H}_5)_3\text{P}]_4\text{Pd}} \longrightarrow \underset{\mathbf{13}}{[\text{PE-PPh}_2]_4\text{Pd}} + (\text{C}_6\text{H}_5)_3\text{P} \qquad (12)$$

We have briefly explored the possibility that a polymeric ligand could be better than an electronically similar ligand for metal complexation by comparing the efficacy of polyethylene carboxylate versus propanoate in complexation of the ruthenium cluster $[(\text{RO}_2)_6\text{Ru}_3\text{O(H}_2\text{O)}_3]^+\text{RO}_2^-$.[14] This heptacarboxylate triruthenium cluster was used as an alcohol oxidation catalyst by Drago.[15] As discussed below, we have found that this ruthenium cluster catalyst can be attached to polyethylene and reused. During the course of these studies we attempted to use the simple exchange process illustrated in equation 13 to prepare a polyethylene-bound analog of this Ru_3 cluster catalyst. While this reaction was successful based on visible spectroscopy (the deep green solution was water-white after the polyethylene carboxylate ligated cluster had precipitated), ICP analysis of the filtrate showed that some Ru remained in solution. The amount of Ru present in the first filtrate corresponded to ca. 9% of the charged Ru_3 cluster. However, while some of this Ru may be the in the form of the unreacted propanoate cluster, some of the Ru may also have been due to contamination of the initial Ru_3 cluster preparation since a second precipitation recovered > 98 % of the Ru. Thus, our preliminary work suggests that these polymer-bound ligands may be better ligands even in comparison to electronically equivalent low molecular weight analogs.

$$PE\text{-}CO_2H \ + \ (CH_3CH_2CO_2)_6Ru_3O^+ \cdot CH_3CH_2CO_2^- \ \rightleftharpoons$$

$$(PE\text{-}CO_2H)_1(CH_3CH_2CO_2)_5Ru_3O^+ \ + \ CH_3CH_2CO_2H \ \rightleftharpoons \qquad (13)$$

$$(PE\text{-}CO_2H)_2(CH_3CH_2CO_2)_4Ru_3O^+ \ + \ CH_3CH_2CO_2H \ \overset{\text{repeatedly}}{\rightleftharpoons}$$

$$(PE\text{-}CO_2H)_6Ru_3O^+ \ + \ CH_3CH_2CO_2H$$

14

Rhodium(I) Catalyzed Hydrogenation Reactions

While we have used the terminally functionalized ethylene oligomers as ligands for a variety of catalysts, the results obtained in studies of hydrogenation catalysts analogous to Wilkinson's catalyst serve as the most useful comparison of these polymer catalysts versus a standard homogeneous catalyst or other insoluble polymer-bound catalysts.[13] The hydrogenation catalyst **10** is compared with $ClRh(P(C_6H_5)_3)_3$ in hydrogenations of various alkenes in Figure 1. Also on this Figure are data for hydrogenations carried out using either a commercial sample of

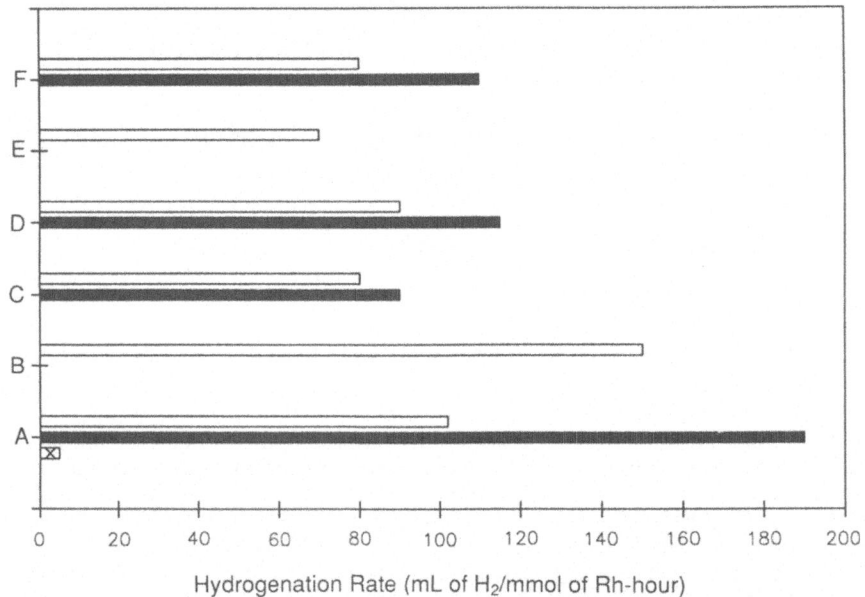

Hydrogenation Rate (mL of H_2/mmol of Rh-hour)

Figure 1. Hydrogenation rates of various alkenes using catalyst **10** (open bar), $ClRh[P(C_6H_5)_3]_3$ (filled-in bar) or DVB-crosslinked polystyrene supported $ClRh[P(C_6H_5)_3]_3$ (cross-hatched bar). The alkenes used were; A = 1-octene; B = styrene; C = α-methyl styrene; D = cyclooctene; E = cyclododecene; F = Δ^2-cholestene. Hydrogenation reactions were carried out at 100 °C in *para*-xylene using 10% catalyst.

DVB-crosslinked polystyrene supported ClRh(P(C$_6$H$_5$)$_3$)$_3$ analog as a catalyst. As can be easily seen, the homogeneous polyethylene-bound catalyst is comparable in activity to ClRh(P(C$_6$H$_5$)$_3$)$_3$. The differences in reactivity are most likely due to the electronic differences between the alkyldiphenylphosphine ligands in **10** versus the triphenylphosphine ligands in ClRh(P(C$_6$H$_5$)$_3$)$_3$. The data in this figure also show that the polystyrene-bound catalyst has significantly lower activity than either ClRh(P(C$_6$H$_5$)$_3$)$_3$ or **10** under comparable conditions. In addition, the heterogeneous polystyrene-bound catalyst's activity reportedly is further diminished with larger substrates (e.g. Δ^2-cholestene) while the relative activity of **10** versus ClRh(P(C$_6$H$_5$)$_3$)$_3$ is the same for this alkene as it is for α-methylstyrene (activity of **10**/activity of ClRh(P(C$_6$H$_5$)$_3$)$_3$ = 0.78 for these two alkenes).[16]

As noted above, preparation of **10** could be accomplished either starting with the phosphine-free chlororhodium-ethylene dimer or with a phosphine containing rhodium catalyst. In the latter case, the catalyst activity gradually changed as the catalyst was recycled, usually rising to a constant value within 3-4 dissolution-precipitation cycles. A typical example of this behavior which we ascribe to gradual replacement of the triphenylphosphines in ClRh[P(C$_6$H$_5$)$_3$]$_3$ by the oligomeric phosphine **4** is shown in Figure 2.

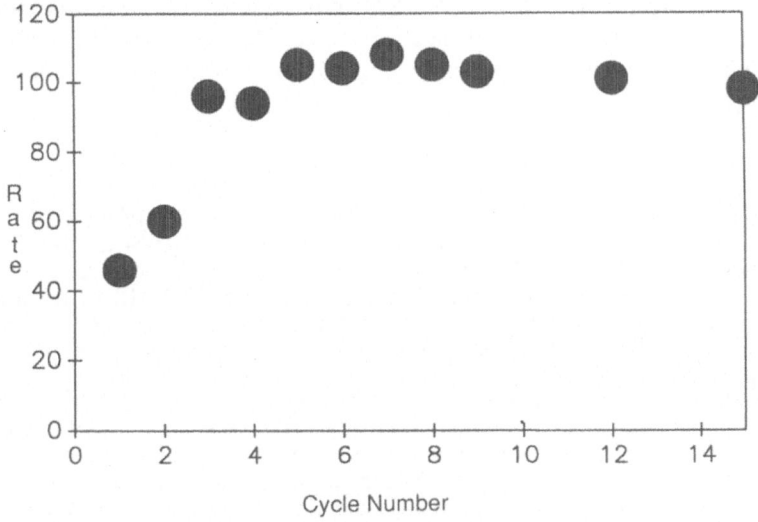

Figure 2. Activity of **10** prepared from ClRh[P(C$_6$H$_5$)$_3$]$_3$ and **4** increase as a function of the extent of recycling of the catalyst. Catalytic activities were measured as initial rates at 100 °C in toluene or xylenes.

One concern in using polymer-bound catalysts generally and these polyethylene-bound catalysts in particular at 100 °C is their chemical and thermal stability. The recyclability of homogeneous catalysts is not always known and examples exist in which *ortho*-metalation or formation of bulk metal or metal clusters occurs.[17] Since some darkening of the solutions occurred during repeated us of **10**, this question was of immediate concern to us. We studied the stability of **10** in several ways. First, the reactivity of **10** through some eighteen cycles of 1-octene hydrogenation was relatively constant (Figure 3). Second, the relative rates of hydrogenation of alkenes relative to 1-octene using fresh **10** and "used" **10** (which had been through 10-12 cycles of alkene hydrogenation) were the same. Third, ^{31}P NMR spectra of solutions of fresh **10** and used **10** were essentially the same. Both samples contained minor amounts of the oxide of phosphine **4**. However, the rest of the

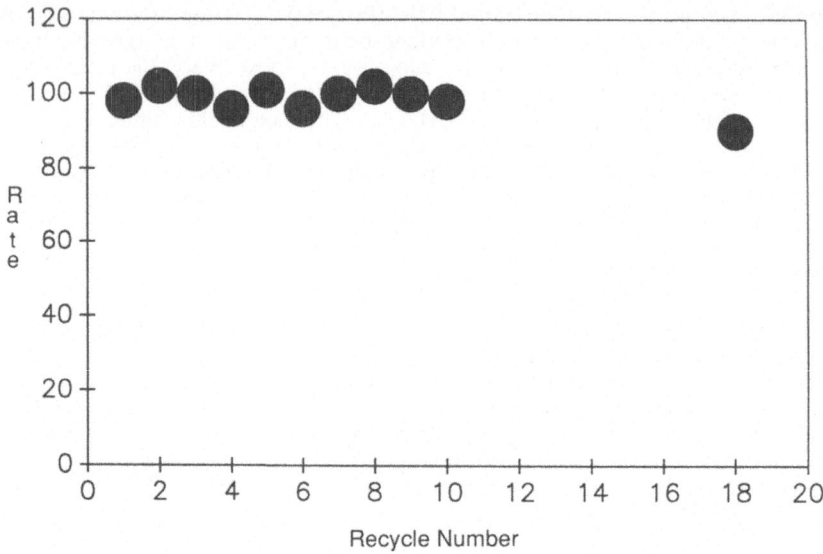

Figure 3. Recyclability of **10** in 1-octene hydrogenation in xylenes at 100 °C. After 18 cycles, the catalyst **10** had lost about 10% of its activity. The total turnover (moles of 1-octene hydrogenated/mole Rh catalyst was ca. 218. However, the catalyst was successfully used in other subsequent hydrogenations as well so this number is a minimal value.

phosphine ligand **4** was coordinated to Rh based on the absence of a signal for free **4**. Control experiments using $ClRh(P(C_6H_5)_3)_3$ and excess $P(C_6H_5)_3$ at 100 °C showed that a separate signal for free **4** would have been expected if extensive decomposition of **10** to bulk Rh metal had occurred. Fourth, solid state ^{31}P NMR of fresh and used catalysts were similar. This experiment further supports the conclusions reached from the solution-state ^{31}P NMR studies. Finally, a "three-phase" test like that used by Collman provided positive evidence that the catalyst was homogeneous.[18] In this test, the observation that the polyethylene-bound hydrogenation catalyst both as a "freshly prepared" and as a "used" (10-cycles) catalyst hydrogenated a divinylbenzene-crosslinked polystyrene-bound vinyl group indicated that the active catalyst was a homogeneous species. While the hydrogenation rate in this case was significantly slower than that seen using $ClRh(P(C_6H_5)_3)_3$ and the same insoluble polystyrene-bound alkene substrate, a heterogeneous Rh/C catalyst is unable to effect this hydrogenation.

The original goal of preparing a homogeneous recyclable hydrogenation catalyst was met using **10** as evidenced by the data in Figures 2 and 3 and by the results discussed above. A further consideration was that little or no rhodium be lost during recovery by leaching. The absence of any significant rhodium in the filtrate was confirmed by inductively coupled plasma atomic emission spectroscopic analysis. After digesting any residue obtained on evaporation of the filtrate with $HClO_4$ and HNO_3, a solution was obtained which was analyzed and compared with NBS atomic absorption standards. The results showed < 0.1 % of the charged rhodium was lost in the first reaction cycle of a 1-octene hydrogenation.

Nickel(0) Catalyzed Cyclooligomerization Reactions

While hydrogenations of alkenes using **10** were successful in demonstrating that polyethylene ligands could be useful as recoverable analogs of homogeneous catalysts, it was also important to demonstrate that these polymeric ligands can be used like their low molecular weight counterparts to usefully modify a catalysts' activity. We used the known Ni(0) catalyzed cyclooligomerization of butadiene (equation 14) to

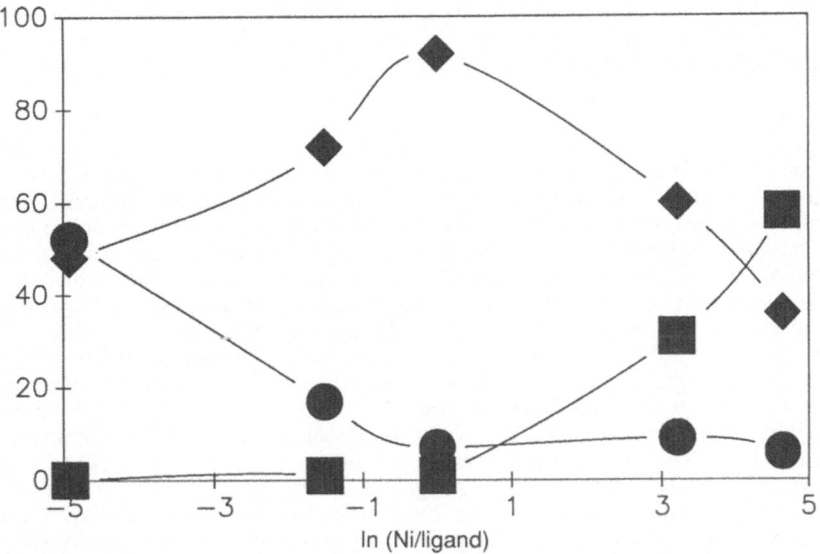

$$(14)$$

demonstrate this.[19,20] In this nickel catalyst chemistry, hindered triarylphosphites combined with nickel(0)-alkene complexes are highly active but selective catalysts. Depending on the ratio of ligand:nickel and on the structure of the phosphite, the predominant product of the cyclooligomerization can be 4-vinylcyclohexene, 1,5-cyclooctadiene or (E,E,E)-1,5,9-cyclododecatriene (cf. Figure 5 and 6).[21]

Figure 5. Distribution of cyclooligomerization products as a funciton of the Ni to ligand ratio in the cyclooligomerization of 1,3-butadiene using di-(*ortho*-tolyl)-phosphite functionalized ethylene oligomers in toluene at 100 °C: filled circles, 4-vinylcyclohexene; squares, (E,E,E)-1,5,9-cyclododecatriene; and diamonds, 1,5-cyclooctadiene.

Comparison of the data in Figures 5 and 6 shows that these polyethylene-bound nickel(0) catalysts behave in every respect like mimics of their low-molecular weight phosphite-ligated homogeneous analogs. In addition to having essentially idential phosphite ligand concentration-dependent product selectivity, these polyethylene-bound nickel(0) diene cyclooligomerization catalysts also had activities which paralleled known nickel(0) chemistry. The highest activities for diene cyclooligomerization were seen with hindered triarylphosphites.

The procedures used in these cyclooligomerizations are typical of those used in other examples of homogeneous catalysis with these polyethylene-ligated metal complexes. For example, using the polyethylene aryldi(*o*-tolyl)phosphite as a ligand

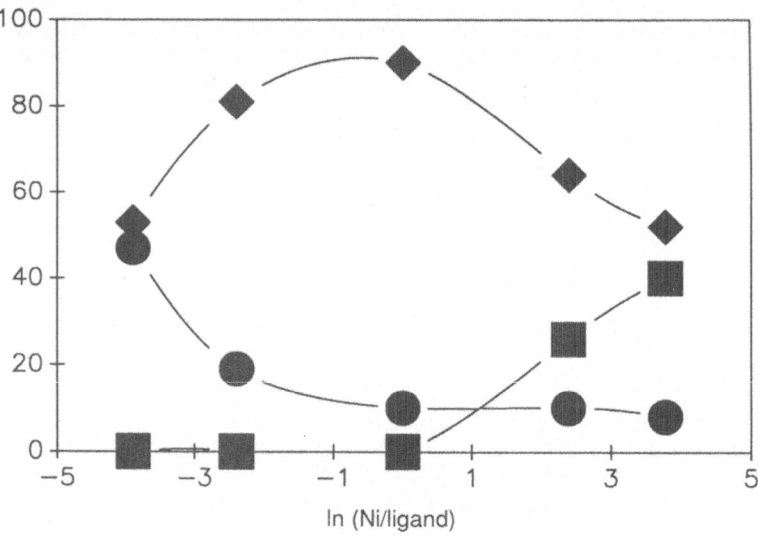

Figure 6. Distribution of cyclooligomerization products as a funciton of the Ni to ligand ratio in the cyclooligomerization of 1,3-butadiene using nonadecylphenyl di-(*ortho*-tolyl)phosphite in toluene at 100 °C: circles, 4-vinylcyclohexene; squares, (*E,E,E*-1,5,9-cyclododecatriene; and diamonds, 1,5-cyclooctadiene.

butadiene can be cyclooligomerized to produce 1,5-cyclooctadiene as a major product as follows. Initially the ligand (0.5 g, 0.25 mmol) and starting nickel(II) catalyst precursor (Ni(acac)$_2$, 0.06 g, 0.24 mmol) was weighed into a pressure bottle equipped with a magnetic stirring bar. then this reaction apparatus was connected to a pressure manifold and purged several times with N$_2$ and vacuum. Addition of 30 mL of freshly distilled toluene (distilled from sodium-benzophenone to remove H$_2$O and O$_2$) was followed by two freeze-pump-thaw cycles. Heating the resulting suspension to 90 °C then produced a light green solution. Cooling this solution to room temperature produced a light green polymer powder. Butadiene (3 g) was condensed into the reaction vessel after it had been further cooled to -20 °C. Then a solution of triethylaluminum in hexane (0.8 mL, 0.8 mmol) was added and the reaction mixture heated to 95 °C. The resulting light yellow solution was visually identical to a solution prepared using triphenylphosphite as a ligand. After the reaction was complete, the catalyst was quantitatively recovered by simply cooling the reaction mixture to precipitate the polyethylene. Centrifugation and decanting the supernatant away from the polyethylene-entrapped Ni(0) catalyst followed by washing yielded a polyethylene powder containing active catalyst which was reusable to cyclooligomerization of butadiene on heating and dissolution.

Multi-step Oxidation Reduction Reactions

The use of polymer-bound catalysts in multi-step reactions was one of the first recognized advantages of "heterogenized" homogeneous catalysts.[3] The general idea was that two insoluble supports would effectively isolate two otherwise reactive species because they would be unable to react with one another. This concept, which already had precedent in Nature and in the use of mixed strong acid-strong base ion exchange resins, was shown to be feasible in a number of reactions. For example, Pittman's group explored a number of such reactions in which the product of one catalytic reaction was the substrate for the second. Thus, his group was able to bind a diene oligomerization catalyst and hydrogenation catalysts to two insoluble resins and thus effect the overall transformation shown in equation 15.[22] Organic applications of this technique also were developed. An excellent example of such applications can be

$$(15)$$

found in Patchnorik's "Wolf and Lamb" reactions in which a strong base and electrophile are rendered compatible by each polymer bound reagent's insolubility (equation 16).[23]

$$(16)$$

In the course of developing the diene cyclooligomerization and hydrogenation catalysts described above, we recognized that soluble polymers could also be used in such multi-step processes. However, unlike the case with two different polystyrene-bound species, the use of one soluble and one insoluble polymer bound species would enable the facile isolation of the product *and* each catalyst or reagent. Flotation techniques can be used to separate two insoluble polymers from one another. However, separation of a soluble polymer from an insoluble one would be simpler since it would only involve a filtration step. Kinetic isolation of each polymer-bound species would be possible even with one soluble polymer-bound species because of the diffusional constraints associated with soluble macromolecules.

We chose the general reaction scheme shown in equation 17 to examine the feasibility of kinetically isolating a soluble polymeric reagent or catalyst from an

$$(17)$$

insoluble one. In this scheme, we envisioned using H_2 as the penultimate reductant in the presence of a transition metal catalyst like **10** and a stoichiometric oxidant. Preliminary experiments showed that H_2 was kinetically slow to react with various oxidants including a polyvinylpyridinium-bound chlorochromate (PVPCC). This Cr(VI) oxidant is a polymeric oxidant developed by Frechet's group which is commercially available and which can be washed extensively with water such that no soluble chromium species leach from the resin.[24] The absence of any soluble chromium species was deemed to be essential to avoid unwanted decomposition of **10**.

The simultaneous oxidation-reduction shown in equation 17 was subsequently carried out in two ways. First, the unsaturated alcohol, 10 mole % of the oligomer catalyst and a four- to six-fold excess of PVPCC were all suspended in toluene and then heated to 100 °C in a H_2-atmosphere. Under these conditions, H_2-uptake began as soon as the catalyst dissolved and the overall process took 12-24 hours. Hot filtration at the reaction's end separated the spent and unused chromium oxidant from the hydrogenation catalyst and carbonyl product. Subsequent cooling precipitated the hydrogenation catalyst which was recovered and reused. The resulting solution contained the saturated carbonyl compound. Alternatively, the oxidation was initially performed at 70 °C under conditions where the polyethylene-bound rhodium catalyst was insoluble and inactive. Then the reaction mixture was heated to 100 °C under hydrogen to effect hydrogenation. Isolation of the chromium reagent, catalyst and product was similar to the above procedure. However, the catalyst 10 was recycled successfully more times under the latter procedure. Examples of the applicability of this oxidation-reduction reaction sequence are shown in Table 1 below.[25]

Table 1. Oxidation/Reduction of Unsaturated Alcohols at 70 °C and 100 °C.[a]

Starting Material	Time (h)	Product	Yield (%)
3-cyclohexenemethanol	24/24	cyclohexanecarboxaldehyde	85
10-undecen-1-ol	24/24	undecanal	90
1-phenyl-10-undecen-1-ol	20/20	1-phenylundecan-1-one	95
11-dodecen-2-ol	24/20	2-decanone	80
3-cyclohexenylphenylmethanol	24/20	cyclohexyl phenyl ketone	95
3-cyclohexenylphenylmethanol	20/24	3-cyclohexenyl phenyl ketone	95[b]

[a]The oxidation step was carried out initially at 70 °C and heating to 100 °C was delayed until the oxidation was complete. Simultaneous oxidation/reduction was also feasible if the entire reaction sequence was carried out at 100 °C. [b]The hydrogenation catalyst used was $ClRh(P(Ph)_3)_3$. No hydrogen uptake occurred presumably because of oxidation of the phosphine ligand.

Control experiments using $ClRh[P(C_6H_5)_3]_3$ in place of 10 showed that the presence of an oligomeric ligand like 4 was essential to this reaction's success. Ordinary homogeneous catalysts and/or their phosphine ligands are too susceptible to oxidation by the resin bound Cr(VI) to be useful in reactions like equation 17.

Palladium(0) Catalyzed Allylic Substitution Reactions

Palladium catalysis using $Pd(P(C_6H_5)_3)_4$ as a homogeneous catalyst is a very useful reaction in allylic substitution reactions in which both carbon and heteroatom nucleophiles substitute for allylic carboxylate groups.[26] These sorts of reactions catalyzed by $Pd[P(C_6H_5)_3]_4$ and other reactions carried out using or $[P(C_6H_5)_3]_2Pd(O_2CR)_2$ complexes can be similarly carried out using diphenylphosphine terminated polyethylene as a ligand to make the catalyst recoverable and recyclable.[27] In these examples, the catalyst was synthesized as described in equation 12. In this reaction, the orangish solution of the tetakis(triphenylphosphine)palladium(0) complex yielded a water-white solution of triphenylphosphine after being heated to 100 °C with 4 and then cooled to 25 °C. The resulting polyethylene-bound palladium(0) catalyst was then used in reaction 17 to form allylic amines from allylic acetates. Yields and typical examples of this reaction are listed in Table 2.

$$+ \quad \text{(18)}$$

Table 2. Palladium(0)-Catalyzed Reaction of Allyl Benzoate with Piperidine.[a]

Time (min)	Cycle	Yield (%)
10	1	100
10	2	95
20	3	90
20	4	72
20	5	68

[a]The reaction was carried out at 100 °C using 2% $(PE\text{-}PPh_2)_4Pd$ as a catalyst in toluene.

Ruthenium-Catalyzed Oxidation of Alcohols with Oxygen

Oxidation catalysts are of general interest especially when the penultimate oxidant is oxygen. When such catalysts are used as homogeneous catalysts, procedures which permit their quantitative separation from products, recovery and reuse can be useful. While many variations of insoluble polymers have been described which are useful for this purpose, there are comparatively few examples of such systems being used for oxidation catalysts.[15,28] Further, metal clusters, which can exhibit especially interesting activity, have not been as routinely incorporated into insoluble polymers such that they are used and recovered without decomposition. We have found soluble polyethylene ligands usefully complex oxidation catalysts based on triruthenium cluster catalysts recently described by Drago. Specifically, we have found that addition of polyethylene oligomers containing a terminal carboxylic acid group to solutions of $(CH_3CO_2)_6Ru_3O(H_2O)_3$ allows us to quantitatively recover and reuse this ruthenium cluster as shown in equation 13 above.[14]

We have successfully used this polyethylene-bound cluster catalyst to homogeneously catalyzed oxidation of both primary and secondary alcohols (equation 19). In the case of the primary alcohols, the oxidation stopped at the aldehyde stage. Representative examples of these oxidations are listed in Table 3.

Some initial oxidations were carried out in toluene solution. However, in these cases, there was a significant amount of oxidation of the solvent. Switching to chlorobenzene avoided this problem without affecting catalyst activity.

$$\text{R} \overset{\displaystyle \text{HO}}{\underset{\displaystyle \text{R'}}{\diagdown\diagup}} \quad \xrightarrow[\text{O}_2,\ 100\ ^\circ\text{C}]{(RCO_2)_6Ru_3O(H_2O)_3{}^+ \quad O_2CR^-} \quad \text{R} \overset{\displaystyle O}{\underset{\displaystyle \text{R'}}{\diagdown\diagup}} \qquad (19)$$

R = alkyl or aryl; R' = alkyl, aryl or H

Table 3. Alcohol Oxidation Using O_2 and Polyethylene Bound Ru_3 Clusters.[a]

Alcohol	Substrate/Catalyst	Turnovers (d^{-1})	Time (h)
1-dodecanol	265	25	120
1-dodecanol	624	26[b]	62
1-dodecanol	886	16[c]	69
		13[d]	40
		20[e]	88
cyclohexanol	116	8	42
geraniol	136	30[f]	17

[a]The oxidation reaction was carried out at 100 °C in toluene or (preferably) chlorobenzene using magnetic stirring for agitation and 40 psi of O_2. [b]A Vibromixer was used in place of of the magnetic stirrer. [c]This reaction was carried out in an autoclave under 100 psi of O_2. [d]This reaction was carried out in an autoclave using recycled catalyst. [e]This reaction was carried out in an autoclave using twice recycled catalyst. [f]This reaction used octane solvent.

An unresolved problem with these polyethylene-bound ruthenium clusters is their thermal decomposition. This decomposition, which was noted by Drago, was confirmed by UV-visible studies of solutions of these polyethylene-bound catalysts and is the principle limit to catalyst recyclability in the case of this cluster catalyst.

Summary

Soluble functionalized polymers of the sort described here are equivalent to conventional ligands for homogeneous catalysts in terms of a catalyst's activity and selectivity. However, using linear functional polymers as macromolecular ligands provides significant advantages in several ways. These advantages have been illustrated in this paper with the specific example of polyethylene derived ligands. These functionalized polyethylenes have the same solubility as high molecular weight polyethylene. These ligands and their metal complexes are insoluble at room temperature but dissolve in several solvents at elevated temperature to produce homogeneous solutions of catalysts. Catalyst recovery and separation from reaction products simply requires cooling and filtration. The physicochemical properties of the polyethylene-like tail of these catalyst ligands is also useful in other ways. For example, the diffusional limitations associated with these macromolecules have been used to kinetically isolate two otherwise incompatible reagents. Soluble polymeric ligands also have some advantages over insoluble polymers in that they are easier to synthesize and characterize.

Acknowledgment

Support of our research in polyethylene chemistry by the National Science Foundation and of our research in catalysis and ligand synthesis by the Petroleum Research Fund and the Robert A. Welch Foundation is gratefully acknowledged.

References

1. R. B. Merrifield, *Angew. Chem. Int. Ed. Engl.* **1985**, *24*, 799-810, and references therein; C. Birr, "Aspects of the Merrifield Peptide Synthesis", Springer-Verlag: Berlin, 1978.

2. F. R. Hartley, "Supported Metal Complexes: A New Generation of Catalysts", D. Reidel: Amsterdam, 1985; D. C. Bailey and S. H. Langer, *Chem. Rev.* **1981**, *81*, 109-148.

3. D. D. Whitehurst, *CHEMTECH,* January 1978, 44-49; W. T. Ford, *CHEMTECH*, April 1987, 246-250; D. E. Bergbreiter, *CHEMTECH*, November 1987, 686-690.

4. N. K. Mathur, R. E. Narang and R. E. Williams, "Polymers as Aids in Organic Chemistry", Academic Press: New York, 1980; "Polymer Supported Reactions in Organic Synthesis"; P. Hodge and D. C. Sherrington, Eds.; Wiley: New York, 1980.

5. V. N. R. Pillai and M. Mutter, *Acc. Chem. Res.* **1981**, *14,* 122-130.

6. V. Schürig and E. Bäyer, *CHEMTECH*, April 1976, 212-214.

7. D. E. Bergbreiter, Z. Chen and H.-P. Hu, *Macromolecules* **1984**, *17*, 2111-2116.

8. D. E. Bergbreiter, M. D. Hein and H.-P. Hu, *Macromolecules* **1988**, *21*, in press.

9. P. N. Young, R. P. Quirk and L. J. Fetters, *Adv. Polym. Sci.* **1984**, *56*, 1-90.

10. J. H. Rosedale and F. S. Bates, *J. Am. Chem. Soc.* **1988**, *110*, 3542-3545.

11. H. Rachapudy, G. G. Smith, V. R. Raju and W. W. Graessley, *J. Polym. Sci., Polym. Phys. Ed.* **1979**, *17*, 1211.

12. D. E. Bergbreiter, H.-P. Hu, J. R. Blanton, M. D. Hein, K. Huang, R. Chandran and D. R. Treadwell, submitted for publication

13. D. E. Bergbreiter and R. Chandran, *J. Am. Chem. Soc.* **1987**, *109*, 174-179.

14. D. E. Bergbreiter and D. R. Treadwell, submitted for publication.

15. C. Bilgrien, S. Davis and R. S. Drago, *J. Am. Chem. Soc.*, **109**, 3786-3787 (1987).

16. R. H. Grubbs, L. C. Kroll and E. M. S. Sweet, *J. Macromol. Sci., Chem. A.* **1973**, *7*, 1047-1063.

17. J. P. Collman, L. S. Hegedus, J. R. Norton and R. G. Finke, "Principles and Applications of Organometallic Chemistry, 2nd Edition", University Science Books: Mill Valley, CA, 1987.

18. J. P. Collman, K. M. Kosydar, M. Bressan, W. Lamanna and T. Garret, *J. Am. Chem. Soc.* **1984**, *106*, 2569-2579.

19. P. W. Jolly and G. Wilke, "The Organic Chemistry of Nickel", Academic Press: New York, 1975.

20.   D. E. Bergbreiter and R. Chandran, *J. Org. Chem.* **1986**, *51*, 4754-4760.

21.   P. Heimbach and H. Schenkluhn, *Top. in Curr. Chem.* **1980**, *92*, 45-108.

22.   C. U. Pittman, Jr. and L. R. Smith, *J. Am. Chem. Soc.* **1975**, *97*, 1749-1754.

23. A. Patchornik, *Nouv. J. Chim.* **1982**, *6*, 639-643.

24.   J. M. J. Frechet, P. Darling and M. J. Farrall, *J. Org. Chem.* **1981**, *46*, 1728-1730.

25.   D. E. Bergbreiter and R. Chandran, *J. Am. Chem. Soc.* **1985**, *107*, 492-4793.

26.   J. Tsuji, "Organic Synthesis with Palladium Compounds", Springer Verlag: Berlin, 1980.

27.   D. E. Bergbreiter and D. Weatherford, submitted for publication.

28.   J. J. Spivey, *Ind Eng. Chem. Res.*, **26**, 2165-2180 (1987); C. L. Bailey and R. S. Drago, *Coo Chem. Rev.*, **79**, 321-332 (1987).

AUTOXIDATION CATALYSTS BOUND TO POLYMER COLLOIDS

Warren T. Ford*, Rama S. Chandran, M. Hassanein[†], and Hayrettin Turk

Department of Chemistry
Oklahoma State University
Stillwater, Oklahoma 74078

INTRODUCTION

Transition metal catalysts can be classified as either heterogeneous or homogeneous.[1,2] Heterogeneous catalysts are usually finely divided metal or metal oxides supported on carbon, silica, alumina, a zeolite or some other inorganic solid. They are used at high temperatures in continuous processes with contact times of reactants with catalysts of seconds to minutes. Their activity is generally high because of the high temperature employed. Homogeneous catalysts[1,3] are specifically synthesized metal compounds that are used in solution at temperatures ranging from subambient to perhaps 200 °C. Homogeneous catalysts are generally more selective, due both to the lower operating temperatures and to their specifically designed structures. Thus heterogeneous catalysts are usually considered for large scale, rapid reactions, and homogeneous catalysts for smaller scale reactions that require higher chemical selectivity.

Catalysts also can be classified as chemical or biological. Chemists have devoted great effort in recent years to mimic enzymes, with substantial progress but without yet coming close to the goal of a synthetic catalyst that has both the activity and the specificity of an enzyme. Since enzymes have high specificity and perform at ambient temperature, they are usually described as homogeneous catalysts, and their kinetics are analyzed by the methods of homogeneous solution kinetics.[4] However, in living organisms enzymes usually perform in cytoplasmic gels, in membranes, and in other macromolecular assemblies within the cells of living organisms. Even our digestive enzymes, which have been studied so extensively, perform heterogeneously. Industrial processes that use immobilized enzymes also are heterogeneous reactions.[5]

We are exploring synthetic transition metal catalysts in the heterogeneous aqueous environments of polymer colloids. Potential uses of such catalysts are decontamination of water and industrial organic chemical syntheses. Initial efforts are focused on oxidations. Organic pollutants in water can be difficult to degrade if they lack polar functional groups. The oxidation of hydrocarbons to alcohols or carbonyl compounds should make them more easily degraded by the normal biological waste treatment methods. Colloidal polymers are promising as catalyst supports because of their high surface areas. Heterogenized homogeneous catalysts supported on colloidal particles 10-1000 nm in diameter should have activities much greater than those supported on 0.1-5 mm particles. For a strictly mass transfer limited reaction, the activity of a heterogeneous catalyst is inversely proportional to its particle radius.[6]

[†] Permanent address: Department of Chemistry, Faculty of Science, Tanta University, Tanta, Egypt.

Colloids are either charge stabilized by ionic functional groups or sterically stabilized by pendant polymer chains on their surfaces.[7] Coagulation of charged colloids is prevented by coulombic repulsion. Stability depends on both the net surface charge and the composition of the medium in which the particles are suspended. In water dissolved electrolytes screen the coulombic repulsions, such that charged polymer colloids usually coagulate at critical concentrations of uniunivalent electrolytes in the range 0.1-0.5 M. Sterically stabilized colloids have well-solvated polymer random coils on their surfaces. Stabilization is due to the large loss of entropy that would result from interpenetration of polymer coils of different particles. Their stability too depends on the medium in which they are suspended. They coagulate when a non-solvent for the surface-bound polymer causes contraction of the random coils.

The major goals of our catalysis research are to determine the range of catalytic processes that may be performed in colloidal environments and to understand the mechanisms of chemical reactions that take place at the surface and within colloidal particles. Despite the wide use of colloidal polymers (also known as latexes) in rubber, coatings, and adhesives, few chemical reactions have been carried out with them. The likely reasons are that their structures are not well understood, and they can be difficult to maintain in the colloidal state without coagulation. These problems require that we investigate not only chemical reactions, but also colloid stabilization and surface structures, for chemical reactions using active colloidal catalysts proceed at the particle surface.

Colloidal catalysts are different from homogeneous, heterogeneous, and enzyme catalysts. We aim with colloidal catalysts to take advantage of the specificity of soluble transition metal compounds by immobilizing them on high surface area organic polymer supports in free-flowing aqueous dispersions. They resemble heterogeneous catalysts in that the chemical reactions take place at a phase interface. However, the liquid phase is water, and the solid phase is a flexible organic polymer. The particle surface is not that of a rigid, highly ordered crystal, but a disordered array of amphiphilic polymer chains solvated by both organic compounds and water. Having both hydrophilic and lipophilic properties, colloidal catalysts operate in environments more nearly like those of enzymes than those of homogeneous or heterogeneous catalysts.

Colloidal catalysts are related to other new types of catalysts that combine the advantages of homogeneous and heterogeneous catalysts. A few examples of colloidal metal particles stabilized by adsorbed organic polymers have been investigated as hydrogenation catalysts.[8] Analogues of homogeneous transition metal complexes have been supported on macroscopic particles of polymers and silica to take advantage of the activity of homogeneous catalysts and the recoverability of heterogeneous catalysts.[9,10] By electrocatalysis oxidation-reduction reactions are performed at the surfaces of electrodes.[11] Photocatalysis can occur at the surfaces of colloidal semiconductor particles.[12] Catalysts bound to synthetic membrane bilayer and mutilayer vesicles, micellar catalysts, and catalysts in microemulsions are designed to have the activity of homogeneous catalysts with hopes for enzyme-like specificity conferred by their heterogeneous environments.[13,14] All of these new types of catalysts work at the interfaces of condensed phases. The catalytic environments of charged colloidal catalysts most nearly resemble those of micelles, membranes, and water-soluble polyelectrolytes[15] because of the high charge density.

This chapter describes our initial investigations of cobalt complexes bound to colloidal polymers for the autoxidations (in which dioxygen is the stoichiometric oxidant) of tetralin, 2,6-di-tert-butylphenol, and 1-decanethiol. The reactions of these water-insoluble organic compounds proceed faster in the colloids that they do with the same cobalt complexes as catalysts in aqueous solution. Autoxidations have been chosen because of their potential for low cost use for the decontamination of water and for chemical manufacturing processes.

AUTOXIDATION OF TETRALIN[16]

A standard method of oxidation of alkylaromatic hydrocarbons is treatment with dioxygen in acetic acid with cobalt(II) as catalyst and HBr as promoter.[17] 1,2,3,4-Tetrahydronaphthalene (tetralin) yields 1,2,3,4-tetrahydro-1-naphthol (1-tetrol) and 3,4-dihydro-1(2H)-naphthalenone (1-tetralone) as the major products and smaller amounts of 1-hydroperoxy-1,2,3,4-tetrahydronaphthalene (THP).[18-22] We reasoned that copolymer latexes containing acrylic acid or methacrylic acid repeat units might bind Co(II) and also serve to concentrate tetralin from an aqueous dispersion into the organic polymer particles near the active sites.

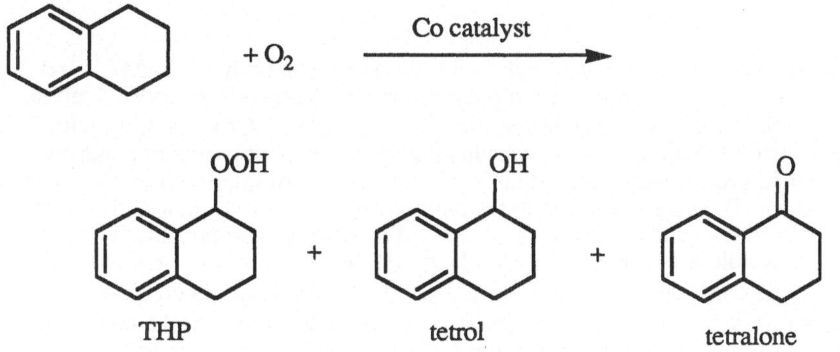

$$\tag{1}$$

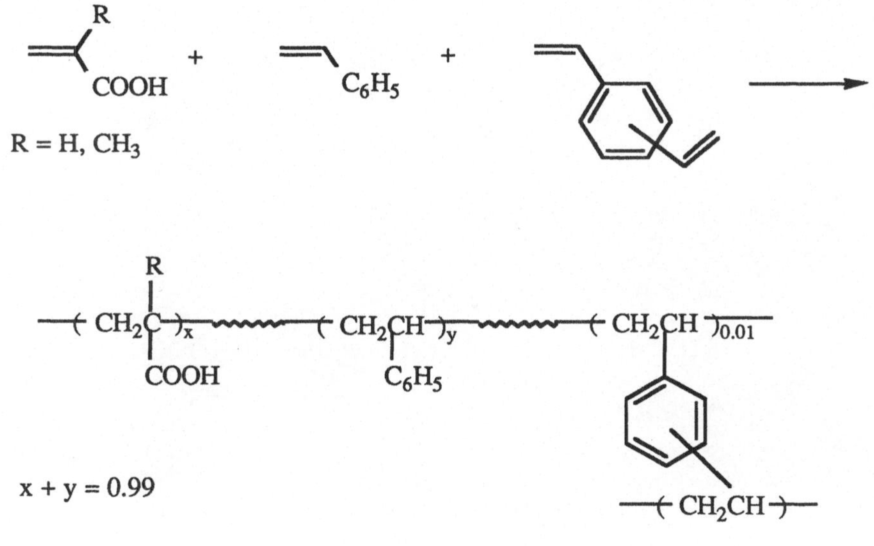

THP   tetrol   tetralone

R = H, CH$_3$

$$\tag{2}$$

x + y = 0.99

161

Table 1. Colloidal Acrylic Copolymers and Catalysts.

Catalyst	Copolymer [a]	COOH/ Co(II)	mg-atom Co/ g polymer	$d,$[b] nm
RC-1	79:20 MA:1	18.6	0.11	58
RC-2	79:20 MA:1	8.0	0.25	58
RC-3	79:20 AA:1	17.4	0.12	62
RC-4	23:76 AA:1	2.3	4.2	140

[a] Mole % styrene, acrylic acid (AA) or methacrylic acid (MA), and divinylbenzene (55 % technical grade) in the monomer mixture.
[b] Number average particle diameter measured on transmission electron micrographs.

Copolymer latexes of styrene, acrylic acid (AA) or methacrylic acid (MA), and divinylbenzene were prepared by emulsion polymerization using sodium dodecylsulfate as the surfactant and $K_2S_2O_8$/NaHSO$_3$ as a redox initiator (eq. 2).[23,24] Cross-linking with divinylbenzene insured that the polymer was insoluble during all subsequent experiments. The carboxylic acid groups were converted first to the potassium salt and then to Co(II) salts by addition of cobalt(II) acetate solution to the copolymer latex with agitation in an ultrasonic bath to produce the latex catalysts listed in Table 1. The latexes were purified by ultrafiltration through a 0.1 μm cellulose acetate/nitrate membrane (Millipore) until the conductivity of the filtrate at 25 °C was constant at 40×10^{-6} ohm^{-1} cm^{-1}. Purified latexes contained 1-2% solids. An upper limit of 3×10^{-6} for the fraction of Co(II) not bound to the latex was established by addition of 1,10-phenanthroline to the ultrafiltrate and UV-visible spectrophotometric analysis of the Co(II) complex. By the same criterion, addition of 6 mol of pyridine per mol of Co(II) to form the active catalysts did not extract cobalt ions from the latex. Thus practically all of the Co(II) was bound to latex. A typical transmission electron micrograph of catalyst RC-1 is shown in Figure 1.

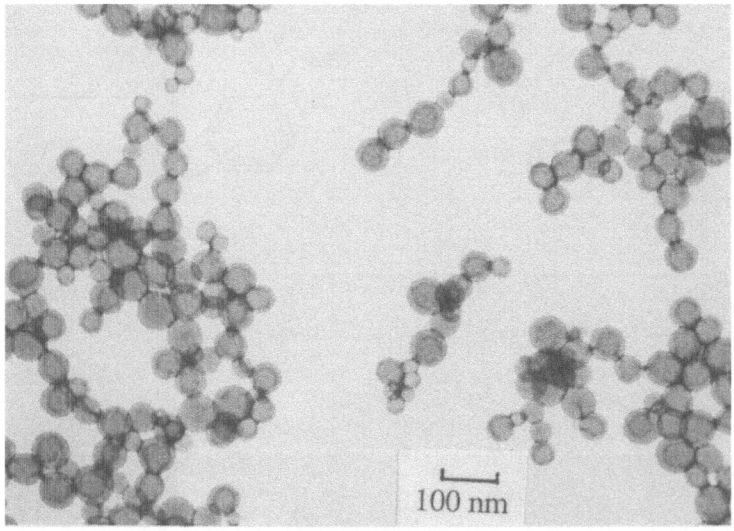

Figure 1. Transmission Electron Micrograph of Catalyst RC-1

Table 2. Oxidation of Tetralin Using Latex Catalysts.[a]

Catalyst[b]	Co(II)	Py	Tetralin	Tetralin	Tetrol	Tetralone	THP
	------Initial concns, mM-------			--------------Products formed[c]-------------			
RC-1	0.66	4.0	61	35	22	43	3
RC-2	0.66	4.0	61	42	21	37	2
RC-3	1.40	8.4	130	37	20	43	2
RC-4	1.40	8.4	130	55	16	27	1
RC-1	1.90	12.0	100	38	21	41	3
CoPy[d]	1.90	12.0	100	57	15	28	2
CoAc[e]	1.90	--	110	76	2.5	23	1

[a]Reactions were carried out at 50 ± 0.1 °C for 24 h and O_2 pressure of 720 mmHg (ca. 20 mmHg less than atmospheric pressure) with magnetic stirring of 30.0 mL of reaction mixture. Catalysts were equilibrated at 50 °C under oxygen prior to addition of tetralin. [b]Table 1. [c]From GLC analysis of organic compounds extracted after coagulation of the latex catalyst with sodium chloride solution. [d]Catalyst was soluble Co(II)-pyridine in water at pH 8. [e]Cobalt(II) acetate in acetic acid.

Autoxidations of tetralin using the latex catalysts are listed in Table 2. The reactions were followed by the consumption of dioxygen using a gas buret. A typical plot of volume consumed vs. time is shown in Figure 2. Even though the solubility of tetralin in water is estimated to be less than 10^{-4} M, oxidation of more than 60% of 0.06-0.13 M dispersed tetralin in the aqueous colloids occurs in 24 h at 50 °C at one atm pressure of dioxygen. The rate of tetralin oxidation with a colloidal CoPy catalyst is twice as fast as with aqueous CoPy and nine times as fast as with cobalt(II) acetate in acetic acid, as shown in Figure 2. An amine ligand such as pyridine was necessary for Co autoxidation activity in water. The final reaction mixtures were analyzed by gas chromatography. Only tetralin, tetrol, tetralone, and THP were found in significant amounts. Pure THP was prepared by independent synthesis and found to decompose partially to tetrol and tetralone under the analytical GC conditions. The product distributions in Table 2 are corrected for THP decomposition.

A complete mechanism for the autoxidation of alkylaromatic hydrocarbons by cobalt(II) in acetic acid has not been established,[25,26] although a complex rate law has been determined for tetralin.[18-22] The reaction most likely proceeds by a free radical chain mechanism in which the purpose of the cobalt ions is to provide a high steady state concentration of free radicals by catalysis of the decomposition of THP. The free radical nature of the autoxidation of tetralin with the colloidal CoPy catalysts is supported by experiments which showed inhibition of the reaction by 2,6-di-tert-butylphenol and 2,6-di-tert-butyl-4-methylphenol, and by a shortening of the induction period and increase of the reaction rate when azobis(isobutyronitrile) was added to the reaction mixture as a free radical initiator.

The rate of tetralin autoxidation in the aqueous colloid is not dramatically higher than it is in aqueous solution. The increased rate can be explained simply by the ability of the styrene-acrylate and styrene-methacrylate copolymers to absorb tetralin, increasing the reactant concentration near the catalytic sites. The exact environment at the active sites is likely important to catalytic activity, since latexes with the same copolymer composition had different activities, but the contribution of surface composition has not been explored systematically yet.

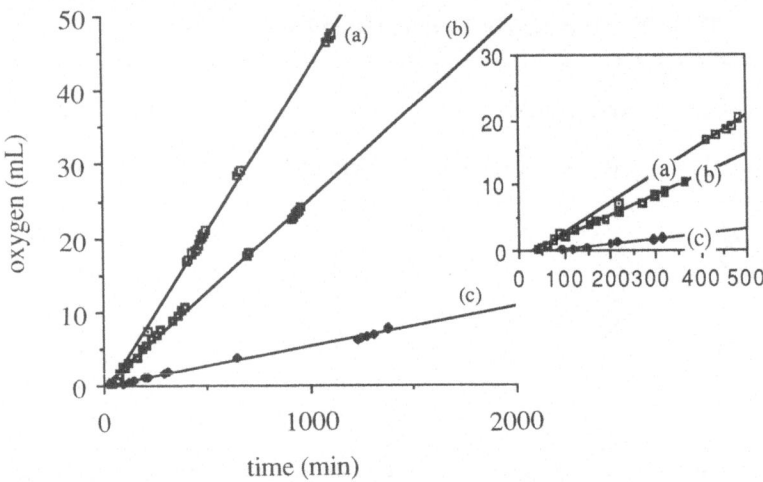

Figure 2. Oxygen consumption with time during autoxidation of tetralin catalyzed by (a) RC-1, (b) aqueous CoPy, and (c) cobalt(II) in acetic acid. Inset figure shows induction periods. Reproduced from ref. 16 by permission of The Royal Society of Chemistry.

AUTOXIDATION OF 2,6-DI-*tert*-BUTYLPHENOL[27]

A variety of transition metal complexes, especially those of cobalt and copper, catalyze the oxidation of phenols. Autoxidations of 2,6-disubstituted phenols in organic solvents produce mainly the 2,6-disubstituted-1,4-benzoquinone and the 3,5,3',5'-tetrasubstituted-4,4'-diphenoquinone with Co catalysts[28] and certain copper catalysts (eq. 3).[29,30] Poly(2,6-dimethyl-1,4-phenylene oxide) is the usual product of autoxidation of 2,6-dimethylphenol with copper-amine catalysts.[31] We have found an efficient autoxidation of 2,6-di-*tert*-butylphenol (**1**, eq. 3) in water catalyzed by CoPcTs bound to colloidal ion exchange resins. Cobalt phthalocyanine derivatives have been bound covalently to several other types of polymers for catalysis.[32,33]

The colloidal polymers were prepared by emulsion copolymerization of chloromethylstyrene (a 70:30 *m:p* mixture) and divinylbenzene. Treatment with trimethylamine converted the chloromethyl groups to quaternary ammonium chlorides. If the quaternized latexes did not contain DVB, they would be water-soluble polyelectrolytes. Addition of an aqueous solution of CoPcTsNa$_4$ to the colloidal ion exchange resins resulted in complete binding of the CoPcTs, when the amount of sulfonate ions did not exceed the amount of quaternary ammonium ions. Ultrafiltration of the colloidal catalysts through 0.1 μm cellulose acetate/nitrate membranes revealed no blue CoPcTs in the filtrates, even though CoPcTsNa$_4$ in water is detectable by human eye to concentrations of $<10^{-7}$ M. Therefore the heterogeneous catalysts listed in Table 3 contained $<10^{-7}$ M of soluble CoPcTs, and soluble CoPcTs did not contribute significant activity when autoxidations were performed with about 10^{-4} M CoPcTs in the heterogeneous mixtures.

Table 3. Ion Exchange Latexes and CoPcTs Catalysts

catalyst[d]	copolymer composition[a]				DF,[b] N+Cl-	R4N+/Co mol ratio	diam.[c] nm
	CMS	DVB	EVB	S			
HT-1	91.2	4.9	3.9	0.0	0.61	6.7	54
HT-3	98.2	1.0	0.8	0.0	0.87	9.5	74
HT-6	91.3	4.8	3.9	0.0	0.81	8.8	55
HT-8	98.2	1.0	0.8	0.0	0.85	9.3	59
HT-9	18.9	1.0	0.8	79.4	0.21	6.7	60

[a] Mol % chloromethylstyrenes, divinylbenzenes, ethylvinylbenzenes, and styrene in the monomer mixture. The first two copolymerizations used sodium dodecylsulfate as emulsifier. The last three used hexadecyltrimethylammonium bromide. [b] Degree of functionalization, the fraction of polystyrene rings substituted with quaternary ammonium groups, determined by Volhard titration of chloride. [c] Number average particle diameter from measurement of transmission electron micrographs. [d] HT-1 was isolated as coagulated particles and redispersed for use as a catalyst. Other catalysts were handled as latexes containing 4.6-9.4% solids.

CoPcTs

$$\text{1} \qquad\qquad\qquad\qquad \text{2} \qquad\qquad\qquad\qquad \text{3} \qquad\qquad (3)$$

Oxidations of 2,6-di-*tert*-butylphenol (**1**) were performed in vigorously mixed aqueous suspensions under 0.9 atm pressure of dioxygen in round bottom flasks connected to a gas buret. Two methods were used to add the phenol **1** to the reaction mixture. Addition of solid **1** to a mixture of catalyst in water at 70 °C gave fine droplets of melted **1** dispersed in the water. Addition of **1** dissolved in methanol gave droplets containing **1** dispersed in 0.1 volume fraction methanol in water. Some reaction mixtures were stirred rapidly with a magnetic bar. Others were shaken with a platform shaker. Results in Table 4 and zero-order initial kinetics show that the colloidal catalyst in water is ten times as active as soluble CoPcTs under stirred conditions (expts. 6 and 7), that 0.1 volume fraction methanol enhances the rate of oxidation (expts. 1 and 3 and expts. 7 and 8), and that shaking gives faster rates than stirring with soluble catalyst (6 and 10) but not with latex catalyst (expts. 8 and 23). The results indicate that the rates of oxidation are partly limited by mass transport phenomena, in absence of methanol and with magnetic stirring. With all colloidal CoPcTs catalysts listed in Table 3 except HT-9, the only product was 3,5,3',5'-tetra-*tert*-butyl-4,4'-diphenoquinone (**2**). With soluble CoPcTsNa$_4$ as catalyst small amounts of 2,6-di-*tert*-butyl-1,4-benzoquinone (**3**) were formed also.

Table 4. Autoxidations of 2,6-Di-*tert*-butylphenol with CoPcTs Catalysts.[a]

expt.	catalyst	MeOH vol. fr.	mixing method	time h	% conv.[b]
1	soluble	0.0	shaking	24	35[c]
3	soluble	0.1	shaking	24	66[c]
4	HT-1	0.0	shaking	24	100
6	HT-1	0.0	stirring	6.	60
7	soluble	0.0	stirring	6	6
8	soluble	0.1	stirring	6	45
23	soluble	0.1	shaking	6	30
9	HT-1	0.1	stirring	6	100
10	HT-1	0.0	shaking	6	100

[a] All experiments were at 70.0±0.1 °C with 1.20±0.01 mmol of **1**, 6.04 x 10^{-2} mmol (5.0 mol %) of CoPcTs, pH 9.0-9.1 adjusted with 4.4 mmol of AMPSO [2-hydroxy-3-[2-hydroxy-1,1-dimethylethyl)amino]-1-propane-sulfonic acid] buffer. All reaction mixtures had a volume of 150 mL and were carried out at ca. 700 mmHg (ca. 40 mmHg less than atmospheric) pressure of dioxygen. [b] Percent of phenol **1** consumed according to GC analysis. [c] 3-4% yield of quinone **3** was found in addition to diphenoquinone **2**.

Table 5. Temperature and pH Effects on Oxidation of 2,6-Di-*tert*-butylphenol[a]

expt.	pH	temp. °C	time min.	% conv.
11	9.0	70	120	100
13	9.0	24	120	16
14	9.0	70	80	79
15	9.0	50	120	50
16	7.0	70	80	17
17[b]	9.0	70	120	55
18	8.0	70	120	31
20	10.0	70	80	100

[a] All experiments used catalyst HT-1, 0.1 volume fraction methanol, mechanical shaking, and the conditions reported in Table 4 unless noted otherwise. [b] Reaction was run under air with a partial pressure of dioxygen of 150 mmHg.

Data in Table 5 shows the pH (expts. 11, 14, 16, 18 and 20) and temperature (expts. 11, 13, 14 and 15) dependence of the autoxidations of **1** with colloidal catalyst HT-1. We conclude from increased oxidation rates with increasing pH in the 7-10 range that the phenoxide anion is the likely reactive species. The activities of catalysts prepared from the five different latexes in Table 3 were similar: 77-92% oxidation of **1** occurred in 80 minutes.

Gas buret measurements of consumption of dioxygen as the reactions proceed are shown in Figure 3. After an induction period the decrease in volume is linear with time to about 70% conversion of the phenol **1** to diphenoquinone **2**. Therefore the rate law has a zero-order dependence on substrate concentration. The rate of oxidation is slower under air (expt. 17, Table 5) than under oxygen at atmospheric pressure, but the kinetic dependence on dioxygen concentration has not yet been firmly established. The ratio of zero-order rate constants calculated from Figure 3 for oxidation by colloidal and soluble CoPcTs catalysts is 12. The higher activity of the colloidal catalyst is most likely due to one or two factors: 1) The organic polymer concentrates the phenol **1** near the active sites as a neutral species in the polymer phase or as the 2,6-di-*tert*-butylphenoxide ion at the surface of the positively-charged colloid. 2) The polymer affects aggregation of CoPcTs. Previous research found that CoPcTsNa4 aggregates in water,[35-37] and that in cationic polyelectrolyte solutions CoPcTs is more aggregated than it is in water.[38] This suggests that aggregates have higher catalytic activity than monomeric CoPcTs.

The mechanism of autoxidation of 2,6-di-*tert*-butylphenol using CoPcTs catalysts is not known. It is not a free radical chain reaction, for the substrate is an inhibitor. In contrast to autoxidations of **1** catalyzed by soluble cobalt complexes, with colloidal CoPcTs catalysts the product is exclusively the diphenoquinone **2**. This can be explained by formation of **2** via coupling of two 2,6-di-*tert*-butylphenoxy radicals or coupling of a 2,6-di-*tert*-butylphenoxy radical with a 2,6-di-*tert*-butylphenoxide ion. The charged surface of the colloidal particles concentrates reactants and reactive intermediates into a small fraction of the volume of the aqueous dispersion, favoring products from bimolecular reaction steps over products from unimolecular reaction steps. One can write mechanisms for the reaction in which hydrogen peroxide is a by-product. If formed, hydrogen peroxide could either accumulate in the reaction mixture or act as an oxidizing agent. Deliberate addition of hydrogen peroxide to reaction mixtures did not accelerate the rate of oxidation of **1** under dioxygen and did not cause appreciable oxidation in the absence of dioxygen. If hydrogen peroxide was formed and merely accumulated, the stoichiometry of the reaction would require one mol of dioxygen per mol of **1** rather than the observed one mol of dioxygen per two mol of **1**. Therefore we rule out hydrogen peroxide as a contributor to or significant by-product of the reaction. However, at this time we have not identified the exact role of the CoPcTs or the mechanism by which the diphenoquinone **2** is formed.

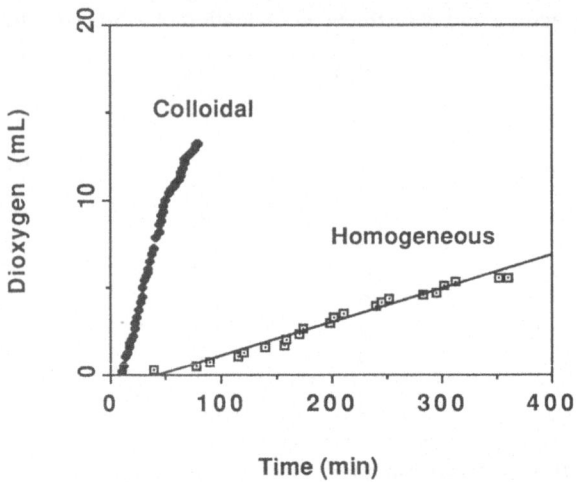

Figure 3. Consumption of oxygen by 2,6-di-*tert*-butylphenol with soluble CoPcTs catalyst (open squares) and with latex catalyst HT-9 (filled squares). Consumption of 0.5 mol O_2 per mol **1** requires 15.02 mL of O_2. Reproduced from ref. 27 by permission of the American Chemical Society.

AUTOXIDATION OF 1-DECANETHIOL[39]

The autoxidation of thiols to disulfides catalyzed by cobalt phthalocyanine derivatives is of great industrial importance for the "sweetening" of petroleum, by processes in which the cobalt phthalocyanine pigment is dispersed in the oil.[40,41] Water soluble polyamines, such as poly(vinylamine), and polyammonium ions, such as ionene polymers, are extremely active cocatalysts for the autoxidation of mercaptoethanol in water at basic pH catalyzed by CoPcTs.[42-51] Cross-linked polyvinylamine and ion exchange resins[52] also serve as cocatalysts, but with lower activity. A major role of a soluble polymeric cocatalyst is to attract both CoPcTs and the conjugate base of mercaptoethanol into the polymeric sub-phase of the clear solution. The polymer also affects the aggregation of the CoPcTs, but the relationship of state of aggregation to catalytic activity is not as well understood.[38] We have found that CoPcTs bound to cationic copolymer latexes is active for autoxidation of 1-decanethiol to the disulfide in aqueous dispersions (eq. 4).

$$4 \ CH_3(CH_2)_9SH + O_2 \rightarrow 2 \ [CH_3(CH_2)_9 \ S]_2 + 2 \ H_2O \qquad (4)$$

The method of latex synthesis for thiol autoxidation catalysts was different from that used for the phenol autoxidation catalysts. Surface active quaternary ammonium ion monomers were used. Latexes were prepared by emulsion copolymerization of 96.2 mol % styrene, 1.0 mol % divinylbenzene (technical 55% active), 0.8 mol % ethylvinylbenzene, and 2.0 mol % of monomer **4**,[53] **5** or **6** with azo(bisisobutyronitrile) as initiator. The conductivity of an aqueous solution of **4** before polymerization was 440 x 10^{-6} ohm^{-1} cm^{-1}. Ultrafiltration of the copolymer latex gave an initial filtrate with condutivity of 20 x 10^{-6} ohm^{-1} cm^{-1} and a

$$CH_2 = \overset{\overset{\displaystyle CH_3}{|}}{C}-COOCH_2CH_2N^+(CH_2)_{17}CH_3 \quad Br^-$$

4

$$\text{(styryl)}-CH_2N^+(CH_3)_2(CH_2)_{15}CH_3 \quad Cl^-$$

5

$$\text{(styryl)}-O-(CH_2)_{12}N^+(CH_3)_3 \quad Br^-$$

6

final filtrate with 6×10^{-6} ohm^{-1} cm^{-1}. Titration of the ultrafiltered latex by the Volhard method detected 96% of the bromide ion initially charged as **4**. Thus the polymer colloids contained an insignificantly small amount of free surfactant. Addition of aqueous CoPcTsNa$_4$ to the latexes completely bound the CoPcTs to form the latex catalysts MH-1, MH-2 and MH-3. Ultrafiltration of each catalyst with a 0.1 µm cellulose acetate/nitrate membrane (Millipore) revealed no blue CoPcTs in the filtrate.

Autoxidation of 1-decanethiol (1.53 mmol) at pH 9.0 in one atmosphere of dioxygen using 600 mg of MH-1 solids in 105 mL of water gave 95% conversion of thiol in 30 minutes at 35°C (expt. 3 in Table 6), even though 1-decanethiol is immiscible with water. The MH-1 contained 0.075 mol % CoPcTs based on 1-decanethiol and 0.1 CoPcTs per quaternary ammonium ion. The amount of thiol consumed was determined by GLC analysis. Slightly more than one mol of O$_2$ per four mols of 1-decanethiol was consumed, which corresponds with the stoichiometry of eq. (4). Results of the oxidations are in Table 6.

After an induction period the volume of oxygen consumed was linear with time to about 65% conversion of 1-decanethiol, which indicates a zero-order rate dependence on 1-decanethiol. The relative zero order rate constants in Table 6 show that the latex catalyst (expt. 3) is eleven times more active than either soluble CoPcTs (expt. 1) or cobalt-free latex (expt. 2). Increased oxidation rates with increasing pH in the range 7.0-9.0 (expts. 3, 4 and 5) indicate that the thiolate ion is the likely reactive species, as found with other polymeric CoPcTs catalysts.[43-51] The increase of rate constants with increasing amounts of latex (at constant loading of CoPcTs in the latex, expts. 3, 6 and 7) and with decreasing amounts of initial thiol (at constant weight and loading of catalyst, expts. 3, 8 and 9) may be attributed to absorption of thiol and thiolate ion into the latexes. With higher initial thiol concentration (expt. 9) there was a decrease in the rate of consumption of dioxygen at >65% conversion. This may be due to accumulation of the didecyldisulfide as well as 1-decanethiol in the polymer. We have not determined yet the kinetic order in dioxygen, but in one experiment (10, Table 6) the rate of oxidation was 0.30 times as fast under air as it was under oxygen (expt. 3) at atmospheric pressure.

Table 6. Autoxidation of 1-Decanethiol with CoPcTs Catalyst MH-1.

expt[a]	latex, mg	CoPcTs, mmol	1-decanethiol mmol	pH[b]	k_{obsd}[c], mL O_2 min^{-1}
1	0	0.0105	1.53	9.0	0.060
2	600	0.0	1.53	9.0	0.034
3	600	0.0105	1.53	9.0	0.68
4	600	0.0105	1.53	8.0	0.44
5	600	0.0105	1.53	7.0	0.129
6	900	0.0158	1.53	9.0	0.72
7	79	0.0014	1.53	9.0	0.30
8	600	0.0105	0.75	9.0	0.37
9	600	0.0105	3.14	9.0	0.60
10	600	0.0105	1.53	9.0	0.20

[a] All experiments were carried out at 35.0 ± 0.1°C and O_2 pressure of 720 mmHg (about 20 mmHg less than atmospheric pressure) with magnetic stirring of 105 mL of reaction mixture. Latex catalysts contained 0.10 mmol Co per mmol quaternary ammonium ions. [b]Adjusted with 4 mL of 0.0125 M $Na_2B_4O_7$ and HCl buffer, except for pH 7 which used Na_2HPO_4 and HCl. [c] Initial zero-order rate constants calculated from the plot of oxygen consumption vs. time.

It is reasonable to ask if surfactant **4** by itself can act as micellar catalyst of autoxidation of 1-decanethiol. In a control experiment, **4** at the same concentration used in the polymer-catalyzed reactions was 0.5 times as active as latex MH-1. However, we have demonstrated that at least 95% of the surfactant in MH-1 is bound to particles, not free in solution, so the contribution of micellar catalysis to the reaction rates of Table 6 is negligible.

Catalyst MH-3, prepared from surfactant **6**, was more active, and catalyst MH-2, prepared from surfactant **5**, was less active than MH-1. We are continuing study of the oxidation kinetics to understand better the mechanism of oxidation. As with soluble poly(vinylamine) and ionene polymers containing CoPcTs, the mechanism does appear to involve the thiolate anion, and the rates of reaction are not simply dependent on the first power of CoPcTs concentration. There is evidence for formation of hydrogen peroxide during the CoPcTs-catalyzed autoxidation of mercaptoethanol.[44,51] At this time we do not know if there are significant differences in the mechanisms of autoxidation of mercaptoethanol catalyzed by polyelectrolytes and autoxidation of 1-decanethiol catalyzed by cationic latexes.

CONCLUSIONS

Aqueous dispersions of charged copolymer latexes are active supports of cobalt catalysts for autoxidations of tetralin by cobalt-pyridine complexes and of 2,6-di-*tert*-butylphenol and 1-decanethiol by CoPcTs. Since these reactants are insoluble in water, a simple explanation of the catalytic activity is that the organic polymer serves to solubilize the reactants in the phase that contains the cobalt catalyst. All three reactions have initial rates that are independent of substrate concentration, as determined by absorption of dioxygen from a gas buret. All three reactions appear to proceed by different mechanisms. Tetralin autoxidation is a free radical chain process promoted by the CoPy complex, whereas the CoPcTs reactions are not free radical chain processes. The thiol autoxidation is reported to involve hydrogen peroxide, whereas the 2,6-di-*tert*-butylphenol autoxidation apparently does not.

Latex-bound transition metal catalysts should be active for a wide variety of reactions of water-insoluble organic compounds in aqueous dispersions. The reactions described here are only a beginning. For practical processes to evolve from this research, methods must be found to recover and reuse the latex catalysts for industrial organic chemical processes, and to coagulate and remove the latexes for water purification processes. The coagulation of charged colloids requires only high electrolyte concentrations. In our experience, some coagulated latexes can be washed to remove adhering salts and redispersed into water to obtain colloids, and other coagulated latexes cannot be redispersed. An understanding of colloid stability and particle dispersion will be required for development of practical processes.

Acknowledgment. This research has been supported by the U. S. Army Research Office and the U. S. Environmental Protection Agency.

References

1. L. L. Hegedus, "Catalyst Design; Progress and Perspectives", Chapter 1, Wiley, New York (1987).
2. B. C. Gates, J. R. Katzer, and G. C. A. Schuit, "Chemistry of Catalytic Processes"; McGraw-Hill, New York (1979).
3. G. W. Parshall, "Homogeneous Catalysis", Wiley, New York (1980).
4. W. P. Jencks, "Catalysis in Chemistry and Enzymology", McGraw-Hill, New York (1969).
5. I. Chibata, T. Tosa, and T. Sato, *J. Mol. Catal.*, **37**, 1 (1986).
6. C. N. Satterfield, "Mass Transfer in Heterogeneous Catalysis", MIT Press, Cambridge, MA (1970).
7. R. H. Ottewill, Chapter 1 in "Emulsion Polymerization", I. Piirma, Ed., Academic Press, New York (1982).
8. H. Hirai, *Makromol. Chem., Suppl.*, **14**, 55 (1985).
9. Y. Chauvin, D. Commereuc, and F. Dawans, *Progr. Polym. Sci.*, **5**, 95 (1977).
10. C. U. Pittman, Jr., in "Comprehensive Organometallic Chemistry"; G. Wilkinson, F. G. A. Stone and E. W. Abel, Eds.; Pergamon Press, Oxford (1983), vol. 8, pp 553-611.
11. C. E. D. Chidsey and R. W. Murray, *Science*, **231**, 25 (1986)
12. M. A. Fox, *NATO ASI Ser. C*, **174**, 363 (1986).
13. J. H. Fendler, "Membrane Mimetic Chemistry"; Wiley, New York (1982), pp. 293-491.
14. J. H. Fendler, *Science*, **223**, 888 (1984).
15. N. Ise, T. Okubo, and S Kunugi, *Acc. Chem. Res.*, **15**, 171 (1982).
16. R. S. Chandran and W. T. Ford, *J. Chem. Soc., Chem. Commun.*, 104 (1988).
17. R. A. Sheldon and J. K. Kochi, "Metal-Catalyzed Oxidations of Organic Compounds", Academic Press, New York (1981), Chapter 3, pp 38-42; Chapter 5, pp 120-127.
18. Y. Kamiya, A. Beaton, A. Lafortune, and K. U. Ingold, *Can. J. Chem.*, **41**, 2020 (1963).
19. A. E. Woodward and R. B. Mesrobian, *J. Am. Chem. Soc.*, **75**, 6189 (1953).
20. Y. Kamiya, *J. Catal.*, **33**, 480 (1974).
21. A. S. Hay and H. S. Blanchard, *Can. J. Chem.*, **43**, 1306 (1965).
22. Y. Kamiya and M. Kashima, *J. Catal.*, **25**, 326 (1972).
23. G. A. Campbell and D. A. Upson, *Macromol. Syn.*, in press.
24. D. C. Blackley and S. A. R. D. Sebastian, *Brit. Polym. J.*, **19**, 25 (1987).
25. S. Lunak, M. Vaskova, P. Lederer, and J. Vepreksiska, *J. Mol. Catal.*, **34**, 321 (1986).
26. E. Baciocchi, L. Mandolini, and C. Rol, *J. Org. Chem.*, **45**, 3906 (1980).
27. H. Turk and W. T. Ford, *J. Org. Chem.*, **53**, 460 (1988).
28. S. A. Bedell and A. E. Martell, *J. Am. Chem. Soc.*, **107**, 7909 (1985).
29. J. P. J. Verlaan, R. Zwiers, and G. Challa, *J. Mol. Catal.*, **19**, 223 (1983).
30. J. P. J. Verlaan, C. E. Koning, and G. Challa, *J. Mol. Catal.* **20**, 203 (1983).
31. A. S. Hay, H. S. Blanchard, G. F. Endres, and J. W. Eustance, *J. Am. Chem. Soc.*, **81**, 6335 (1959).
32. L. Rollmann, *J. Am. Chem. Soc.*, **97**, 2132 (1975).
33. H. Shirai, S. Higaki, K. Hanabusa, Y. Kondo, and N. Hojo, *J. Polym. Sci., Polym. Chem. Ed.*, **22**, 1309 (1984).
34. M. Bernard, W. T. Ford, and T. W. Taylor, *Macromolecules*, **17**, 1812 (1984).

35. Z. A. Schelly, D. J. Harward, P. Hemmes, and E. M. Eyring, *J. Phys. Chem.*, **74**, 3040 (1970).
36. L. C. Gruen and R. J. Blagrove, *Aust. J. Chem.*, **26**, 319 (1973).
37. Y.-C. Yang, J. R. Ward, and R. P. Seiders, *Inorg. Chem.*, **24**, 1765 (1985).
38. J. van Welzen, A. M. van Herk, and A. L. German, *Makromol. Chem.*, **188**, 1923 (1987).
39. M. Hassanein and W. T. Ford, *Macromolecules*, **21**, 525 (1988).
40. A. A. Oswald and T. J. Wallace, in "The Chemistry of Organic Sulphur Compounds", Vol. 2; N. Kharasch and C. Y. Meyers, Eds; Pergamon, London (1966), Chapter 8.
41. R. R. Frame, US Patent, 4,298,463 (1981).
42. N. N. Knudo and N. P. Keier, *Russ. J. Phys. Chem.*, **42**, 707 (1968).
43. J. Zwart, H. C. van der Weide, N. Bröker, C. Rummens, and G. C. A. Schuit , *J. Mol. Catal.*, **3**, 151 (1977-78).
44. J. Schutten and J. Zwart, *J. Mol. Catal.*, **5**, 109 (1979).
45. W. M. Brouwer, P. Piet, and A. L. German, *Polym. Bull.*, **8**, 245 (1982).
46. W. M. Brouwer, P. Piet, and A. L. German, *Polymer Commun.*, **24**, 216 (1983).
47. W. M. Brouwer, P. Piet, and A. L. German, *Makromol. Chem.*, **185**, 363 (1984).
48. W. M. Brouwer, P. Piet, and A. L. German, *J. Mol. Catal.*, **22**, 297 (1984).
49. W. M. Brouwer, P. Piet, and A. L. German, *J. Mol. Catal.*, **29**, 335 (1985).
50. W. M. Brouwer, P. Piet, and A. L. German, *J. Mol Catal.*, **31**, 169 (1985).
51. A. M. van Herk, A. H. J. Tullemans, J. van Welzen, and A. L. German, *J. Mol. Catal.*, **44**, 269 (1988).
52. A. Skorobogaty and T. D. Smith, *J. Mol. Catal.*, **16**, 131 (1982).
53. K. Nagai, Y. Ohishi, H. Inaba, and S. Kudo, *J. Polym. Sci., Polym. Chem. Ed.*, **23**, 1221 (1985).

ORGANIC MATERIALS FOR OPTICAL DATA STORAGE

P.S. Kalanaraman*, James E. Kuder and R. Sidney Jones

Hoechst Celanese Corporation
Summit, N.J. 07901 U.S.A.

INTRODUCTION

Optical storage is the emerging technology for storage, processing and retrieval of vast quantities of data. Although magnetics presently dominate this mass storage market, this position is being challenged by optical media.

The advantages of optical media are severalfold. The physical dimensions of the recorded bit in optical recording are on the order of the wavelength of light used, and this combined with close track spacings leads to storage densities which are about ten times those of magnetics. Because the optical head flies at a relatively large distance from the surface of the recording medium, disks are removable from the drive unit. This is not possible with high density magnetic recording and offers considerable advantage in the distribution of large data bases. Also the results of accelerated aging tests on optical media promise lifetimes of ten years or more, in contrast with magnetic media which need to have data refreshed every two years. Finally, optical disks offer fast random access and a potentially low cost operation.

In optical storage, information is stored, usually in digital form, by using a modulated laser beam to induce a physical or chemical change in a recording medium. These recording media fall into three categories listed below in order of increasing challenge for the media manufacturer:

1. Read Only: used for entertainment, software and database distribution.

2. Write Once Read Many (WORM): suitable for personal archiving and mass data storage.

3. Erasable: aimed as a competing technology for magnetic media.

Read Only media are already on the scene in the form of laser discs (video), compact discs (audio) and CD-ROM's, while developments are still occurring in the area of WORM and erasable media. It is the latter two categories to which the remainder of this paper is devoted.

It is appropriate to consider the optical media unit as an integral part of the optical train. As shown in Figure 1, the train starts with a diode laser where near infrared radiation emerges as a diverging oval beam. A collimating lens and anamorphic prism pair effect the conversion into a circular collimated beam having a Gaussian intensity distribution. A polarizing beam splitter cube selects the predominant polarization and is followed by a quarter wave plate which converts linearly polarized light into circularly polarized light. The beam is focused onto the media recording layer and upon reflection, the chirality of the circulalry polarized light is opposite to that of the incident beam. Passage of the reflected beam throught the quarter wave plate converts it back to linearly polarized light, with polarization at right angles to that of the outgoing beam. Because of the rotation of polarization, the beam splitter sends the reflected beam to the side to a photodetector.

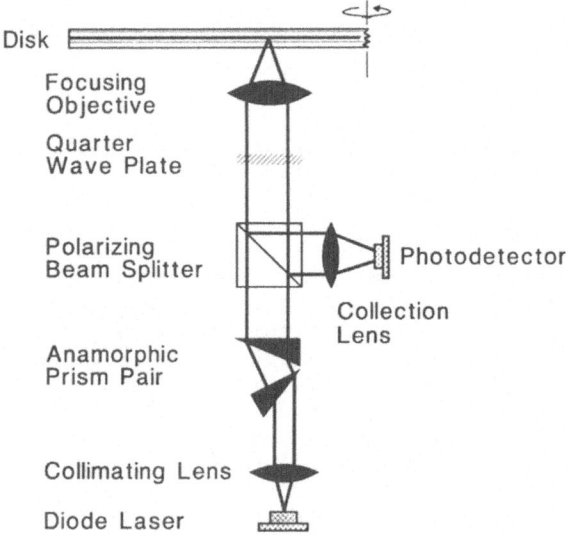

Figure 1. Schematic of optical train.

A diode laser is used because it is compact, inexpensive, reliable and can be directly modulated at appropriate frequencies for data recording. Using the gallium arsenide technology that is currently commercialized, laser power output is limited to tens of milliwatts, and wavelength is confined to 760-850 nm. To maximize data density, the irradiated area is minimized, dictating use of diffraction limited optics. To make the optical design more compact, detection is by reflection, not by transmission, and to allow convenient detection of the reflected beam, polarized light is used to permit separation of the incoming and returning beams. An additional feature is the use of preformatted disks, i.e., disks with tracking, sector and track address information manufactured into the surface of the substrate in the form of grooves and pits.

For random access, a disk which may vary in complexity of structure is the most preferred optical media unit. The substrate provides mechanical strength and may consist of plastic, glass or aluminum. Next, a subbing layer may be added to level microroughness on the surface of the substrate and possibly to protect plastic substrates from solvent-borne coatings (Figure 2). Next, the active (recording) layer is deposited, its thickness being chosen to optimize sensitivity at the laser wavelength by means of a tuned single layer, or further, in a bilayer or trilayer antireflection structure. Finally, an overcoat may be added to protect the active layer from mechanical abuse and chemical attack and to ensure that dust and surface scratches will be out of focus.

In the Air Sandwich design, the active layer is coated onto a transparent substrate and is located within an air gap sandwich structure. Thus, no overcoat is needed. The laser beam is focused through the substrate onto the active layer and consequently the mechanical and optical properties of the substrate play an important part in optical disk construction. The substrate should be highly transparent (¨85%) at the laser wavelength and optically uniform. Optical uniformity not only includes constant, standard path length, but low, uniform birefringence as well. The requirement of low birefringence arises from the use of

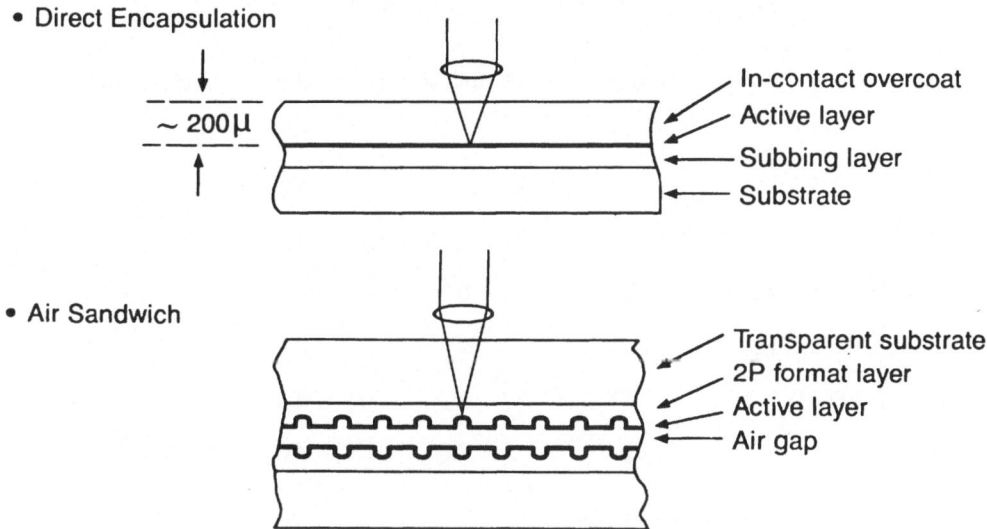

Two examples of disc design. *Top:* in-contact overcoat protects the active layer and ensures that surface defects are out of focus. *Bottom:* transparent substrate serves these purposes.

Figure 2. Schematic drawing of a disk design.

polarized light to address the recording layers. If the substrate has excessive birefringence, the polarization of the reflected read beam is diminished, resulting in decreased intensity at the photodetector. Thus, data detection is impaired. Nonuniform birefringence results in variation of data signal over different areas of the disk. The result is to diminish data reliability and to increase the complexity and cost of the disk drive. In addition, loss of polarization can interfere with laser operation if the depolarized reflected beam optically leaks into the address optical path. The reflected light will in general be out of phase with the cavity modes within the laser and can interfere with them causing the output power to oscillate.

The substrate plays another major role: both tracking and data information, in the form of features, are manufactured into the substrate. These are necessary to make recordable disks practical. Considerable effort has gone into development of molded substrates, satisfying these requirements. Polycarbonate and acrylic resins in combination with injection molding or photopolymerization (2P) process are the most preferred in the manufacture of grooved substrates. In practice, cost has been the dominant factor for most products. Consequently, polycarbonate is used in most mass produced optical disks 130mm diameter or less since yield is relatively high and the required birefringence can be achieved.[1] In the case of 12 inch diameter disks, PMMA is used since molding large polycarbonate substrates with the required optical properties is not yet commericialized. A few products, where performance requirements dominate cost, use glass substrates. Because of the deficiencies of polycarbonate, there is opportunity for new polymers, provided they meet the requirement of competitive cost. A comparison of representative substrate materials is shown in the Table.

TABLE 1   COMPARISON OF SUBSTRATE MATERIALS

PROPERTY	GLASS	POLYCARBONATE	PMMA	EPOXY	OLEFIN
MELT VISCOSITY	NA	0	0	NA	X
BIREFRINGENCE	XX	X	XX	XX	XX
OPTICAL CLARITY	XX	X	XX	XX	XX
SOLVENT RESISTANCE	XX	0	X	XX	X
THERMAL PROPERTIES	XX	X	0	XX	?
MOISTURE REGAIN	XX	X	0	X	XX
COST	0	XX	XX	0	?

KEY: XX - EXCELLENT
         X - GOOD
         0 - POOR
         ? - UNKNOWN
        NA - NOT APPLICABLE

RECORDING LAYER

The recording layer serves to capture the focused laser beam during writing and to consequently undergo some physical or chemical change that alters the intensity of the read beam which is reflected to the detector. The recording layer may itself consist of one or several layers and the functions of light absorption and of marking may reside in a single component or in separate components of the recording layer structure. Light may be absorbed by a metal, dye or pigment film, or a dispersion of metal or pigment particles or a solution of dye in a polymeric binder. The efficiency of light absorption may be enhanced by surface texture[2] or by tuning the thickness d of the recording layer to an antireflection condition, d=1/4n, where 1 is the laser wavelength and n is the refractive index.3 Several material systems with different marking (recording) mechanisms have been explored for this purpose. Some of these mechanisms are summarized below:

1. Deformation, involving a change in surface topography with the formation of pits, bubbles, bumps or the loss of microscopic texture.

2. Phase change, which depends on the difference in refractive indices between crystalline and amorphous phases.4

3. Photochromism, which involves the photochemical conversion of a species absorbing at a first wavelength to another which absorbs at a second wavelength.5

4. The coalescence of discontinuous thin metallic films ("island structure") with a resultant change in reflectivity.6

5. Bilayer alloying, in which reflectivity changes accompany the alloying of adjacent layers.7

6. Magneto-optical alloys that undergo change in the angle of plane polarized light (the Kerr effect) during laser marking.8

As recording media, organic materials have a number of advantages over inorganic materials.9,10 They are less subject to degradation caused by air and moisture. They have lower melting points and thermal conductivities which can lead to higher sensitivity and smaller recorded marks. They can be coated by spin coating techniques, leading to lower fabrication costs as compared to vapor coating required by thin metal films. Finally, the optical and thermomechanical properties of organic media can be tuned by design of molecular structure which is the stock in trade of the synthetic chemist.

A notable difference between inorganic and organic media is that while metals absorb radiation strongly and have high reflectivities at all laser wavelengths, dyes are relatively narrow band absorbers and have lower absorptivities. Thus, the dye must be selected to match its absorption spectrum with the emission wavelength of the laser used. The use of diode lasers emitting in the near infrared (roughly 780-840 nm) for optical recording has led to the design and synthesis of a number of absorbers unknown to classical dye chemists. A representative sampling of these taken from the patent literature is shown in Figure 3.

Following the selection of the dye, it is necessary to decide in what manner the absorber is to be placed on the substrate. With low molecular weight dyes this can be done by vapor deposition, otherwise spincoating from solution is suitable. A problem observed with vapor desposited dyes is that they tend to deposit elsewhere on the disk during writing and cause a decrease in signal to noise ratio (S/N) upon reading. This can be alleviated by overcoating the recording layer, but the overcoat must be chosen so as not to impede the marking process.

A second problem encountered with some dyes is that they tend to crystallize and thus degrade S/N ratio. Both this tendency and the generation of debris upon marking are reduced if the dye is coated together with a polymer to form a dye-binder medium. The presence of covalent or ionic bonds connecting dye and binder has been suggested as a means of preventing segregation of the components of the medium.

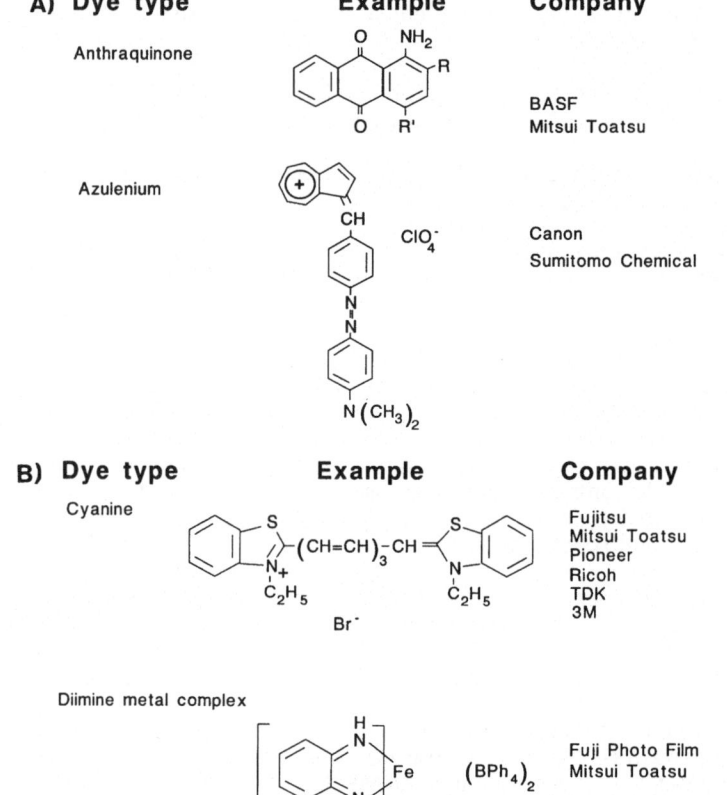

A) Dye type | **Example** | **Company**

Anthraquinone

BASF
Mitsui Toatsu

Azulenium

Canon
Sumitomo Chemical

B) Dye type | **Example** | **Company**

Cyanine

Fujitsu
Mitsui Toatsu
Pioneer
Ricoh
TDK
3M

Diimine metal complex

Fuji Photo Film
Mitsui Toatsu

Figure 3. Some classes of dyes reported as active layer materials in the patent literature.

C) Dye type **Example** **Company**

Dithiol nickel complex

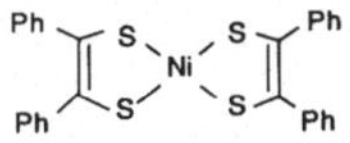

Kodak
Mitsui Toatsu
Ricoh
TDK

Indolizinium

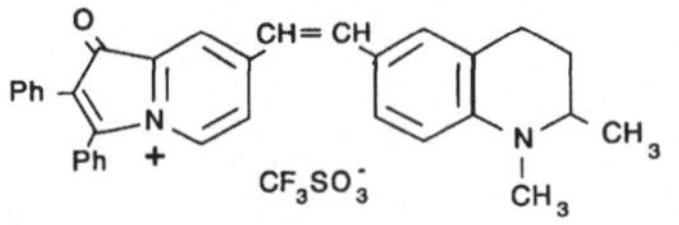

Kodak

D) Dye type **Example** **Company**

Naphthalocyanine

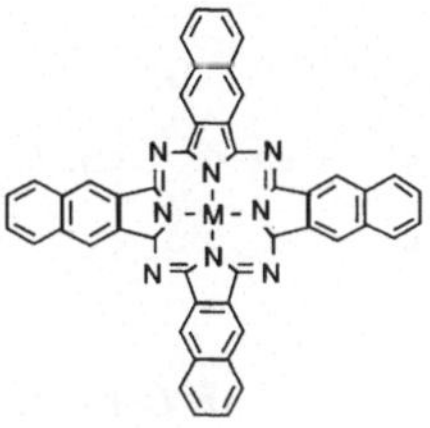

Mitsubishi Chemical
Mitsubishi Paper
Mitsui Toatsu
Xerox
Yamamoto Chemical

Naphthoquinone

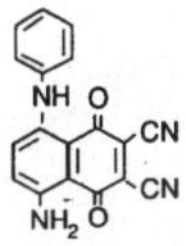

NEC
Sumitomo Chemical

Figure 3 (Continued)

E) Dye type   **Example**   **Company**

Phthalocyanine

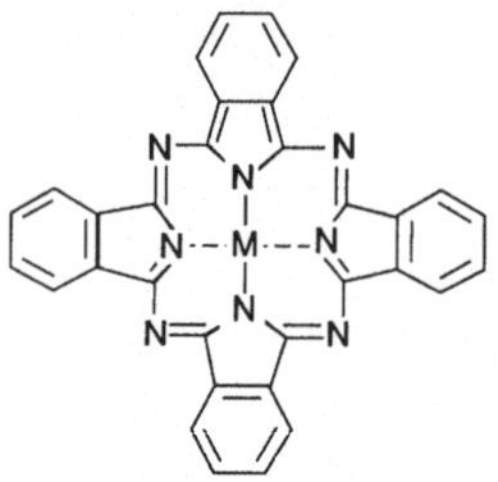

BASF
Canon
Dainichiseika
Hitachi
Philips
Ricoh
TDK
Xerox

Polymethine

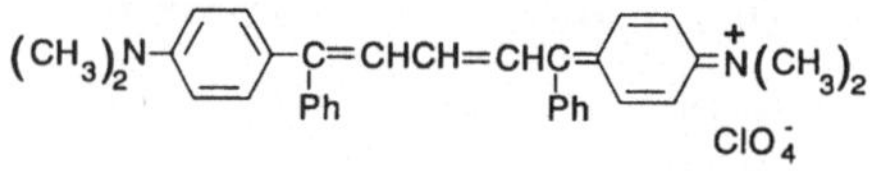

$(CH_3)_2N$——⬡——$\overset{C}{\underset{Ph}{||}}$=CHCH=CHCH=⬡=$\overset{+}{N}(CH_3)_2$

ClO_4^-

Mitsubishi Paper
Philips
Ricoh
3M

F) Dye type   **Example**   **Company**

Pyrylium
Thiapyrylium
Telluropyrylium

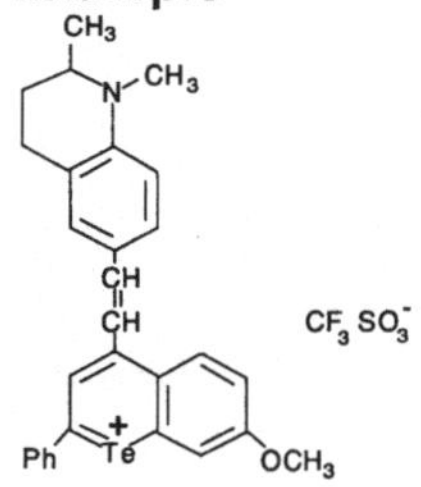

$CF_3SO_3^-$

Kodak
Ricoh
TDK

Squarylium

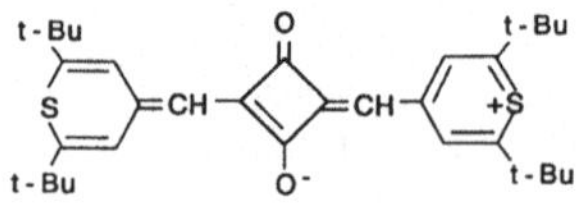

t-Bu

t-Bu

t-Bu

t-Bu

Fuji Xerox
IBM
Philips
Mitsubishi Chemical

G) Dye type   **Example**   **Company**

Triarylammonium

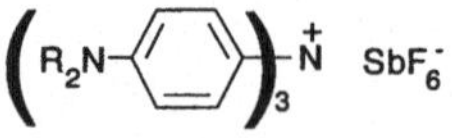

$\left(R_2N-⬡-\right)_3\overset{+}{N}$ SbF_6^-

Sumitomo Chemical

Triquinocyclopropane

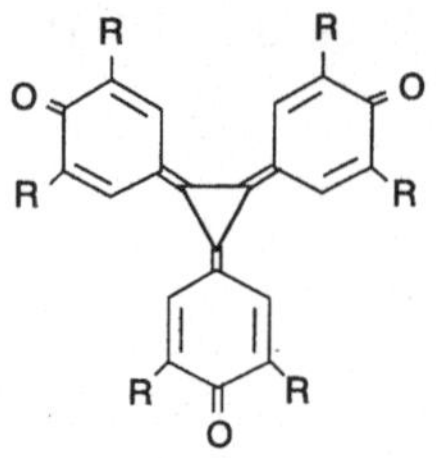

BASF

Figure 3 (Continued)

PIT FORMING MEDIA

Pit formation is the most widely investigated marking process with organic media, in which a pulse of laser light focused onto the active layer results in the formation of a pit surrounded by a rim. If there is a reflective layer under the absorbing layer, then the mark is read as an increase in reflected light intensity; in the absence of a reflecting layer, a decrease in read signal is observed.

A number of attempts have been made to model the pit forming process. Calculations[11] have suggested that the temperature at the center of the focused Gaussian beam may reach as high as 1700oC, well above the decomposition temperature of organic materials. Pits are then formed when the gaseous products displace the surrounding molten material. An alternate model suggests that flow of material driven by surface tension gradients can account for pit formation without invoking thermal decomposition.[12] A third model[13] proposes the formation of a pit by the following sequence: (a) local melting and expansion into a bulge, (b) flow and leveling of the bulge, (c) cooling and contraction to leave a depression surrounded by a rim.

In one study,[14] the same dye was used to form an active layer with a variety of polymers whose glass transition temperatures range from 18 to 190oC. It was found that the threshold for marking was almost completely dominated by the optical density (and thus the efficiency of light absorption) of the film. The toal marking process was found to depend on the molecular weight of the polymer, with partial refilling of the pit occurring with polymers above the critical entanglement molecular weight at which point elastic restoration comes into play.

Another study, however, indicates that the rheological properties of the medium may play a more significant role in the marking process.[15] Here the optical properties of the active layer were kept constant by maintaining a constant active layer thickness and dye-binder ratio (50:50) with a series of ten binders. The latter were chosen so as to have a range of glass transition temperatures from 69o to 310oC and melt viscosities which covered ten decades. Writing sensitivity was found to depend significantly on the melt viscosity of the binder.

Hoechst Celanese Corporation has developed a new class of organic pit-forming media for WORM applications. The performance of a typical example was disclosed recently[16]. This medium could be spincoated on polycarbonate and the recording layer shows high baseline reflectivity (30%) and a threshold energy of approximately 0.4nJ (Figure 4). It is capable of producing CNR of 55dB or greater (Figure 5). The pits are clean with well-defined edges. When read stability, the stability of the read signal after repeated laser reads, was measured, no change is read signal was observed after reading continuously at 0.5mW for 1, 800, 000 cycles (Figure 6). The medium has excellent photolytic stability and showed no significant change in read/write performance after two weeks at 80oC and 95% RH. This combination of excellent environmental and read stability should make the disks outstanding in archivability of data.

Kodak workers have considered the use of thin (approximately 50 nm) ceramic overcoats to suppress the formation of airborne debris during laser writing.[17] The structure of this medium is shown in Figure 7. During recording, heat generated in the dye-binder causes some decomposition of the organic material and softening of the overcoat. Pressure of the gas produced causes the overcoat to form a dome over the evolving pit, while the dome confines the medium and causes it to flow into a rim surrounding the recorded pit. By this approach a medium with good sensitivity

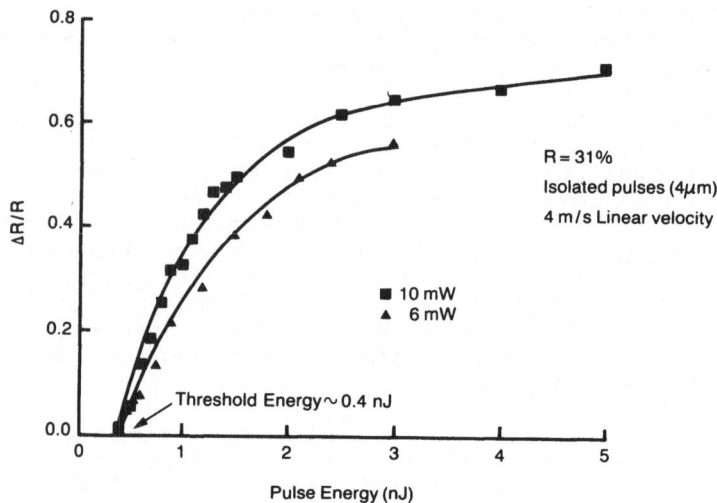

Figure 4. Plot of contrast vs pulse energy for the Hoechst Celanese medium.

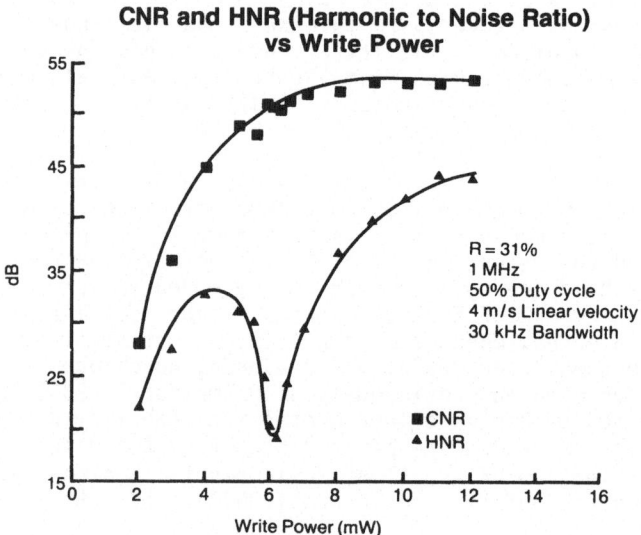

Figure 5. Plot of CNR and HNR vs write power for the Hoechst Celanese medium.

Continuous Read Stability

(0.5 mW Read Power—8 m/s Linear Velocity)

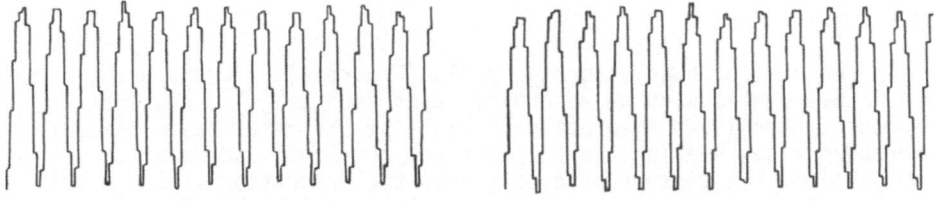

Initial signal   After 1.8 x 10⁶ read cycles

Figure 6. Digitized read signal before and after 1.8 x 106 read cycles.

Overcoat
Dye-binder
Reflector
Smoothing Layer
Substrate

Sensitivity $10mJ/cm^2$

CNR >66dB

BER $10^{-5} - 10^{-6}$

Figure 7. Schematic representation of Kodak's overcoated dye-binder medium. (Gupta, 1984) (approximately 10 mJ/cm2), high CNR (~66 dB) and low BER (10-5-10-6) was obtained.

The functions of absorption and marking in pit forming media can be separated into two separate layers18 as shown in Figure 8. The laser beam is focused through the transparent organic layer onto inorganic layers which control reflectivity and light absorption. The heat generated in the inorganic absorbing layer diffuses to the organic layer and results in a pit which diffracts and scatters the read beam. The organic layer is n-hexatriacontane with a spontaneously aligned layer of zinc stearate between the inorganic and hydrocarbon layers. This antireflection layer is reported to have high sensitivity.

BUBBLE FORMING MEDIA

In the case of bubble forming media (Figure 9), a transparent organic layer is overcoated with an absorbing layer such as gold, platinum or amorphous carbon.19,20 Absorption of light by the metal layer leads to a temperature rise, heating the adjacent polymer layer and causing decomposition. The gaseous products force the separation of the metal layer from the polymer to form a bump in the metal layer. During playback, the laser beam is reflected and diffracted by the bump. When the medium is addressed from the top side, the signal contrast is negative due to diffraction. When the medium is addressed through the transparent substrate, however, the reflectivity at the air-metal interface inside the bubble is greater than that of the unrecorded polymer-metal interface so that the signal contrast is positive. This approach was used in the Thomson-CSF Gigadisc where recording at 14 Mbits/sec was possible at 7 mW.

COLOR FORMING MEDIA

It is possible to do digital recording with media analogous to thermal printing paper.21 The recording mechanism is the thermally induced reaction between a coloring reagent and a phenolic coupling agent which are separated by a light absorbing layer. Light captured by the absorbing layer causes it to melt and allows the two reagents to mix and so produce a colored product. In one embodiment the coloring reagent was Crystal Violet lactone, the coupling agent phenolphthalein and the absorber vanadyl phthalocyanine, each laid down by vacuum deposition. Focused diode laser light produced micron sized spots at a sensitivity of about 150nJ/cm2. A two color medium was also demonstrated using different coloring reagents but a common coupler as shown in Figure 10. Either color could be developed depending on which absorbing layer was focused on.

A variant on this theme is one in which the layers were coated from solution.22 Enhancement of up to 20 dB was claimed for media in which at least one of the layers was deposited as a Langmuir-Blodgett film.

ERASABLE MEDIA

Several erasable media have been reported in the literature including pit-forming media, bubble-forming media and photochromics. The two leading candidates in terms of commercial viability, however, appear to be magneto-optic media and phase change media, both using inorganic materials. Neither one is without drawbacks, however, namely small signal contrast for magneto-optics and long erasure times for phase change media, while environmental stability is still a concern for both. Organic media are less subject to degradation to air and moisture and a few approaches to organic phase change media have been reported. One such approach consists of copper diketonates in a matrix consisting of polystyrene or polycarbonate.23 Recording was conducted with short pulses from a nitrogen

Separate Absorption and Marking Functions

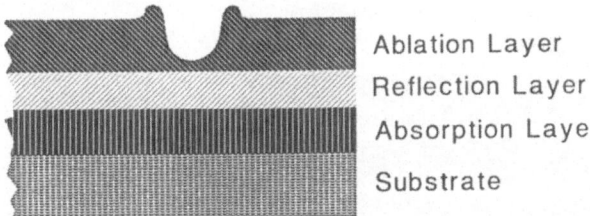

Ablation Layer
Reflection Layer
Absorption Layer
Substrate

Figure 8. Organic pit-forming medium with inorganic reflection/absorption layer. (Mitsuya, 1983)

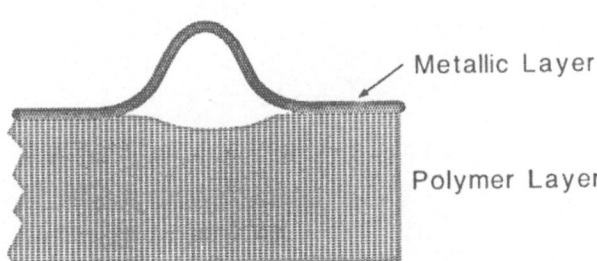

Metallic Layer

Polymer Layer

Figure 9. Schematic of write-once bubble-forming medium.

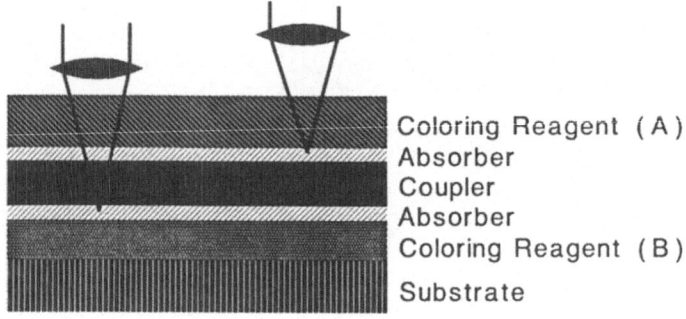

Figure 10. Color forming medium with two different color forming layers.

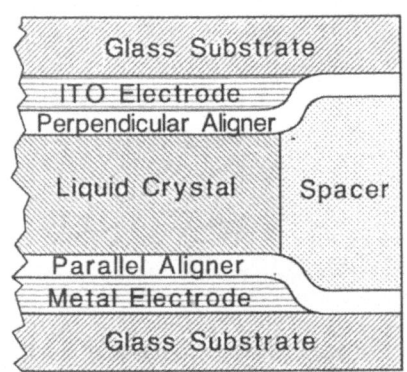

Figure 11. Schematic of liquid crystal medium.

laser, which converted the amorphous material into crystalline domains. Reading was done with a HeNe laser in either the reflective or transmissive mode. Remelting and quenching of the crystalline particles returned the system to its original amorphous structure.

In a related phenomenon, the spectral shifts which accompany aggregation-deaggregation have been reported as a basis of yet another erasable medium.24 The active layer consists of discrete particles of a cocrystalline complex of a pyrylium dye with a polymer such as polycarbonate. Pulsed irradiation of this aggregated film resulted in an instantaneous deaggregation process which was accompanied by a hypsochromic shift of about 100 nm. The recording mechanism is believed to be thermal and a thermal erasure (reaggregation of the dye-polymer complex) was also demonstrated.

Yet another report on an erasable media utilizes the broadening and shifting of an absorption band which occurs upon annealing amorphous solid dyes.25 In the examples reported, writing was done with anthraquinone dyes using microsecond write pulses at 22.5 mW (Kr laser) and millisecond erase pulses at 4 mW. About 100 cycles were possible before failure occurred due to irreversible hole opening.

The use of mesophase transitions in liquid crystal systems has been suggested as an approach to erasable media.26 The embodiment reported is illustrated in Figure 11 and consists of a nematic liquid crystal in an electroded sandwich. The liquid crystal is heated locally by focusing a laser beam onto a metallic absorber which also acts as a reflector for the low intensity read beam. When the write beam is turned off, rapid cooling quenches the medium into a focal conic structure which scatters light. Formation of the scattering centers takes a few milliseconds, so that fast writing is possible but not direct read after write. The scattering centers have circular symmetry so that marks of varying length are not possible. Rather high write powers were reported (80 mW), but high contrast is possible using darkfield optics. Local erasure is possible using laser heating plus a low voltage (20 Vac), while global erasure occurs with a high voltage (70 Vac) with no laser heating. No fatigue was observed after 9000 write/erase cycles.

Polymers with mesogenic side chains are beginning to receive attention as candidate erasable media. As with small molecule liquid crystal analogs, molecular alignment is produced by application of an external electric field. Some examples are shown in Figure 12. Using a fairly simple apparatus, Shibaev and coworkers 27 were able to demonstrate the concept with a polyacrylate having a cyanobiphenyl ether nematic side chain. Laser irradiation of the homeotropically aligned material resulted in scattering centers that were stable below Tg. Similarly, Coles and Simon were able to do laser writing in dyed films of smectic side chain polysiloxane liquid crystals28. Writing was possible in either the homeotropically aligned or scattering textures as a result of material flow along the temperature gradients produced by the laser beam. These marks were stable below Tg but could be erased by either heat or electric field.

A combination of photochemical and thermal effects was used to effect birefringent writing in a polymer having cyanoazobenzene ether mesogenic side chains29. Writing was done with linearly polarized green light (514 nm) and reading of the resulting birefringent pattern was done using 633 nm light whose polarization was at 45o to that of the writing beam. Presumably heating of the film above Tg together with cis-trans isomerization of the azo linkage resulted in a change of molecular orientation which was frozen in upon cooling. Total erasure with no

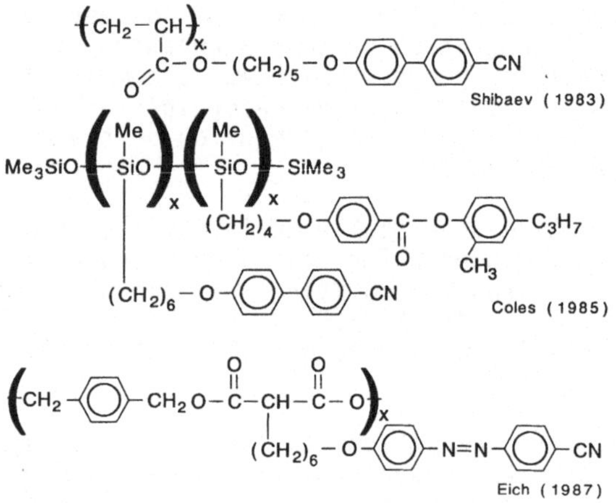

(CH₂–CH)ₓ structures...

$$\left(CH_2-CH\right)_x$$
$$\underset{O}{\overset{\parallel}{C}}-O-(CH_2)_5-O-\!\!\bigcirc\!\!-\!\!\bigcirc\!\!-CN$$

Shibaev (1983)

$$Me_3SiO\left(\underset{\underset{Me}{|}}{SiO}\right)_x\left(\underset{\underset{(CH_2)_4-O-\bigcirc-\underset{O}{\overset{\parallel}{C}}-O-\bigcirc-C_3H_7}{|}}{\overset{|}{SiO}}\right)_x SiMe_3$$

CH₃

$$(CH_2)_6-O-\!\!\bigcirc\!\!-\!\!\bigcirc\!\!-CN$$

Coles (1985)

$$\left(CH_2-\bigcirc-CH_2O-\underset{O}{\overset{\parallel}{C}}-CH-\underset{O}{\overset{\parallel}{C}}-O\right)_x$$
$$(CH_2)_6-O-\bigcirc-N=N-\bigcirc-CN$$

Eich (1987)

Figure 12. Polymeric liquid crystals studied as erasable media.

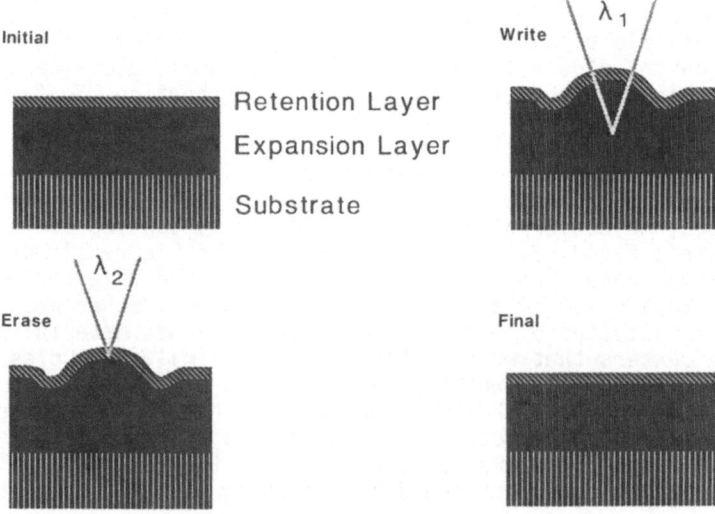

Figure 13. Write and erase in the ODI bilayer bump-forming medium. (Hartman, 1987)

degradation was achieved by heating to the isotropic phase in an electric field.

ERASABLE PIT-FORMING MEDIA

Erasure of pit-forming media was first reported with a dye-polymer medium sensitive to 633 nm light30. It was possible to effect erasure either in a global manner or addressably by moving the laser 10 m out of focus and scanning across the row of pits at 1 mm/sec. Although optical microscopy showed a track left by the scanning beam, the signal contrast remained virtually unchanged after ten write-erase cycles. Similar results have been disclosed for dye-polymer pit-forming media addressable by diode laser wavelengths31,32.

The effect of a ceramic overcoat on a dye-binder depends on the thickness of the overcoat33. As described above, with a thin (50 nm) overcoat, a pit is formed while the overcoat deforms into a dome. The encoded signal could be erased by several scans across the pits by the laser beam at reduced power. With a thicker (300 nm) overcoat, however, formation of a dome is inhibited and molten material flows from the trailing edge of the mark towards the leading edge. This causes the length of the pit to be independent of the recording pulse length. The thicker overcoat aids erasure, and the CW erase beam does not need to be greatly reduced in power to avoid writing while erasing, although 3-4 passes of the erase beam are still required. No degradation in signal was observed after 40,000 write-erase cycles.

BUMP-FORMING MEDIA

A novel approach to erasability using a bilayer medium (illustrated in Figure 13) was recently reported34. The top layer absorbs at 780 nm and has its glass transition temperature between room temperature and 150oC. The bottom layer, whose Tg lies below ambient, absorbs at 840 nm. Irradiation at 840 nm causes the bottom layer (expansion layer) to expand and deform the top layer producing a bump. Upon cooling, the top layer (retention layer) freezes in the bump with the expansion layer under tension. The bump, which serves to diffract the read beam, appears to be caused solely by thermal expansion with no evidence of bubbles or delamination. To erase, the retention layer is heated and softened by irradiation at 780 nm. This permits the expansion layer to pull the surface back down to its original state. CNR greater than 55 dB were reported (CNR: Carrier-to-Noise Ratio).

PHOTOCHROMIC MEDIA

Photochromic media are reversible photochemical reactions, involving molecular rearrangements rather than mass motion and should be capable of high resolution. As a class, however, they frequently are subject to fading (dark back reaction) and fatigue (irreversible side reactions). Ralston35 examined one of the more promising photochromics, the fulgide medium developed by Plessey. For an acceptable change in optical density of 1.0, the exposure required was 1.2 nJ/ m2, while to erase with short wavelength light 50 nJ/ m2 was needed. No fatiguing was observed after 10 write-erase cycles, however, after 100 cycles erasure became impossible. The pulse energies required for write and erase are in agreement with Tomlinson's calculations on photochromics36. His analysis showed that for readout by sensing absorption of the medium, the required write energy per bit is of the same order as or lower than that for common pit forming write-once media. However, to erase at the same rate as the data are written, the erase laser must have a power as much as an order of magnitude greater than that of the write source.

As a final example of reversible optical media, we may mention CuTCNQ which undergoes the following disproportionation reaction:

$$[Cu+TCNQ-]n \ c \ Cux + [TCNQ]x + [Cu+TCNQ-]n-x$$

The forward reaction results in high contrast and can be effected either by gas lasers[37] or diode laser[38]. Fifty write-erase cycles have been demonstrated for this medium using a defocused diode laser for the erase step.

SUMMARY

Organic materials offer certain advantages over thin metal films as optical storage media. The simplicity of processing them offers cost advantages too. Considerable research effort has been going on around the world to develop organic media with suitable properties to satisfy media requirements. Most of the effort has been toward developing WORM media at present, but erasable media development is likely to follow.

REFERENCES

1. Takashima, M., Okada, T., and Fukunishi, S., 1984, Review of the Electrical Communications Laboratories, 32:253-9.

2. Craighead, H.G., and Howard, R.E., 1981, Appl.Phys.Lett.,39:532.

3. Bell, A.E., and Spong, F.W., 1978, IEEE J.Quantum Electron.,14:487.

4. Bell, A.E.,1982, J.Appl.Phys.,53:3438.

5. Heller, H.G., and Langdon, J.R., 1981, J.Chem.Soc., Perkin II, 341.

6. Jipson, V.B., Lynt, N.H., and Graczyk, J.F., 1983, Appl.Phys.Lett.,43:27.

7. Ahn, K.Y., DiStefano, T.H., Herd, S.R., Mazzeo, N.J., and Tu, K.N., 1982, J.Appl.Phys., 53:3777.

8. Mimura, Y., and Imamura, N., 1976, Appl.Phys.Lett., 28:746.

9. Kuder, J.E., 1986, J.Imaging Technol.,12:140.

10. Pearson, J.M. 1986, CRC Critical Rev. Solid State Mater. Sci., 13:1.

11. Wroebel, J.J., Marchant, A.B., and Howe, D.G., 1982, Appl.Phys.Lett., 44:928.

12. Chung, T.S., 1986, J.Appl.Phys., 60:55.

13. Wissbrun, K.F., 1987, J.Appl.Phys., 62:1123.

14. Law, K.Y., and Johnson, G.E., 1983, J.Appl.Phys., 53:4799.

15. Molaire, M.F., 1987, OSA Topical Meeting on Optical Data Storage, Stateline, Nevada, March 11-13, 1987, Abstract ThA4-1, Pages 90-95.

16. Goldberg, H.A., Jones, R.S., Kalyanaraman, P.S., Kohn, R.S., Kuder, J.E., and Nikles, D.E., 1986, SPIE Proceedings, Optical Mass Data Storage, 695:20.

17. Gupta, M.C., 1984, Appl.Optics, 23: 3950.

18. Mitsuya, M., Terao, M., Kaku, T., Shigematsu, K., and Akagai, M., 1983, J.Appl.Phys., 54:3710.

19. Cornet, J.A., 1983, SPIE Proceedings, 420:86.

20. Maffitt, K.N., Robbins, W.B., and Williams, R.F., 1984, U.S. 4,430,659.

21. Morinaka, A., Oikawa, S., and Yamazaki, H., 1983, Appl.Phys.Lett., 43:524.

22. Nishimura, N., Kawada, H., Haruta, M., Hirai, H., Nozotsuki, N., and Nakagiri, T., 1986, Japan Kokai., 61-63939.

23. Ota, K., 1983, Jap., 58-199, 345.

24. Mey, W., 1985, U.S., 4,513,071.

25. Sporer, A.H., 1987, Appl.Optics, 26:1240.

26. Birecki, S., Naberhuis, S., and Kahn F.J., 1983, SPIE Proceedings, 420:194.

27. Shibaev, V.P., Kostromin, S.G., Plate, N.A., Ivanov, S.A., Vetrov, V.Yu., and Yakolev, I.A., 1983, Polymer Comm., 20:264.

28. Coles, H.J. and Simon, R., 1985, Polymer, 26:1801.

29. Eich, M., Wenderoff, J., Reck, B., and Ringsdorf, H., 1987, Makromol.Chem., Rapid Comm., 8:59.

30. Kuroiwa, A., Namba, K., Asami, S., Aoi, T., Takahashi, K., and Nakagawa, S., 1983, A.Appl.Phys., 22:340.

31. Jones, R.S., Kuder, J.E., and Nikles, D.E., 1985, Abstracts of the 1985 Spring Meeting of the Materials Research Society, San Francisco, CA, (April 15-18, 1985), 112.

32. Ito, M., 1985, Japan Kokai., 60-161, 195.

33. Gupta, M.C., and Strome, F., 1986, J.Appl.Phys., 60:2933.

34. Hartman, J.S. and Lind, M.A., 1987, Abstracts OSA Topical Meeting on Optocal Storage, Stateline, NV, (March 11-13, 1987), page 155.

35. Ralston, L.M., 1983, SPIE Procedings, 420:186.

36. Tomlinson, W.J., 1984, Appl.Optics23:3990.

37. Benson, R.C., Hoffman, R.C., Potember, R.S., Bourkoff, E., and Poehler, T.O., 1983, Appl.Phys.Lett., 42:855.

38. Hoshino, H., Matsushita, S. and Samura, H., 1986, J.Appl.Phys.,25:L341.

CHEMICAL AMPLIFICATION IN THE DESIGN

OF RADIATION-SENSITIVE POLYMERS

J. M. J. Fréchet,*
F. M. Houlihan,
F. Bouchard, and
E. Eichler

A. Hult,
R. Allen,
S. MacDonald,
H. Ito, and
C. G. Willson

Department of Chemistry
University of Ottawa
Ottawa, Ontario K1N 9B4
Canada

IBM Research Division
Almaden Research Center
650 Harry Road
San Jose, California 95120-6099

ABSTRACT

Approaches to the design of resist materials which incorporate chemical amplification have been directed toward three different areas:

- Synthesis of polymers which undergo radiation-induced modification of side-chain polarity in a catalytic process.
- Synthesis of thermally depolymerizable polycarbonates designed to provide imaging *via* an acid catalyzed multiple chain-scission process.
- Development of a imagewise film-deposition technique *via* photoinitiated interfacial cationic polymerization.

All three processes rely on the photogeneration of a catalytic amount of acid within a polymer coating which initiates a cascade of chemical reactions. This results in a resist system with high sensitivity.

INTRODUCTION

Much recent research has been focused on the development of new generation resist materials which possess improved sensitivity and enhanced resolution.[1,2] In particular, rapid progress has been made in the chemistry of negative tone resists with the development of a series of ring-halogenated[3] and halomethylated polystyrenes.[4-7] This approach is especially successful in meeting the requirements for high sensitivity and dry etch resistance but does not meet all resolution goals as these materials suffer from the problem of swelling, a problem common to all negative resists that function on the basis of radiation-induced cross-linking.

---

*Department of Chemistry, Cornell University, Ithaca, New York 14853

An interesting approach to resist materials which can form a negative image, but do not function *via* network formation involves materials which undergo changes in polarity when exposed to radiation. A recent example of this approach is poly(p-formyloxystyrene) a, deep-UV resist[2a,8] which undergoes homolysis of an ester group when irradiated to ultimately afford poly(p-hydroxystyrene) and carbon monoxide. Differential dissolution, to yield either a positive or a negative image, is possible due to the large difference in solubility between the exposed phenolic photo products, and the unexposed formyl polymer.

Since swelling during development is not a problem in systems designed to operate on the basis of radiation-induced polarity changes, significant advances in both resolution and ease of processing can be made by further exploitation of this design. The poly(p-formyloxystyrene) deep-UV resist suffers from a relative lack of sensitivity which, when coupled with the low flux available in the deep-UV from classical exposure systems, makes it impractical in a production environment. In order to resolve this sensitivity problem, which is related to the quantum efficiency of the side-chain deprotection reaction, we have studied approaches involving the phenomenon of *chemical amplification*.[9,10] In this approach, the initial photoevent, which is still of limited quantum efficiency, produces a catalyst which mediates a subsequent chemical reaction. The catalyzed reaction causes a change in polymer structure that can be exploited in image formation. In these systems, the quantum efficiency for generating the altered structure is the product of the quantum efficiency for catalyst production and the catalytic chain length. As a result, the overall quantum efficiency can far great be greater than 1.

RESULTS AND DISCUSSION

A simple implementation of this strategy is shown in our deep-UV materials based on polymers containing t-butyl ester or t-butyl carbonate, pendant groups.[9-12]

These materials all undergo acid catalyzed thermolysis and thus, a resist may be formulated by combining such polymers with an onium salt cationic photoinitiator[13] that is known to generate strong acid upon exposure.

UV-exposure of these resist materials only creates a latent image which is characterized by the presence of a local concentration of strong acid within the unchanged t-butyl ester polymer in the exposed areas. Gentle heating of the polymer causes an acid-catalyzed thermolysis with loss of the side-chain t-butyl ester group, and appearance of an image in the exposed areas. Correspondingly, the unexposed areas are unaffected by the low temperature baking step as these regions do not contain acid. Again, this system leads to positive or negative tone imaging as the image tone can be controlled through choice of an appropriate solvent which selectively removes either the imaged areas, with unprotected side chains, or the areas still covered with the starting polymer. Thus, the t-butylester/onium salt system is similar to the poly(p-formloxystyrene) in its final image development step but requires less radiation, as the light is used only to generate a catalytic amount of acid. This acid is not consumed in the thermolysis, but participates in the cleavage of numerous side groups to provide the desired photomultiplication.

It is known that t-butyl ethers are effective protecting groups for the hydroxyl functioninality.[14,15] A common use is the protection of the hydroxyl groups of amino-acids such as serine (structure 4) or tyrosine.[16] The t-butyl group can be removed under conditions that do not affect ester or peptide linkages. On this basis we have prepared structure 5 and demonstrated[17,18] that this polymer functions in a manner analogous to that described above for the carbonate and ester analogs. This chemistry can be further extended to allylic ether structures[18] such as structure 6.

Another interesting early approach to resists incorporating chemical amplification is typified by our development of the polyphthalaldehyde (PPA) resist system.[9,10] This system operates on the basis of radiation-induced depolymerization; radiation causes direct chain cleavage and results in the complete destablization of the polymer chain with release of monomer. While the polymer itself exhibits low radiation sensitivity, incorporating a small amount of a cationic photoinitiator such as an onium salt, results in a large increase in sensitivity. The material is unique in its ability to self-develop but is difficult to use in practice. We have now prepared a series of polycarbonates which, like PPA, are designed to undergo radiation-induced depolymerization. However, to avoid vapor contamination of the exposure tools which may result when the depolymerization starts immediately upon exposure, as is the case for the PPA resist, our design of these materials incorporates a two-step process whereby a latent image is created by exposure to radiation, while self-development only occurs upon subsequent heating.

The polycarbonates are prepared by phase-transfer catalyzed polycondensation of activated carbonyloxy derivatives of tertiary,[19] or other reactive diols,[20] with simple diols such as bisphenol A (Scheme I). This procedure affords materials with molecular weight $Mn = 10$-$50,000$ while only low molecular weight materials resulted from the amine catalyzed condensation of the tertiary diols with the bis-chloroformate of bisphenol A.

These polymers are sensitive to thermolysis and decomposes rapidly and smoothly to bisphenol A, carbon dioxide, and volatile dienes when heated near $200°$. For example polymer 8 affords only benzene, carbon dioxide, and bisphenol A upon heating. As in the case with t-butyl esters and carbonates, this thermolysis reaction is susceptible to acid catalysis and this property forms the basis of our resist design. The tertiary polycarbonate resist is formulated by casting a polymer film from a solution containing a few percent of substances which will produce strong acid upon photolysis. Triarylsulfonium salts are particularly convenient for exposure in the deep-UV and exposure of thin coatings to very small doses of radiation through a mask creates a latent image in the polymer coating. Upon baking after exposure, development of a visible image is instantaneous as carbon dioxide and volatile dienes are evolved. Since bisphenol A is not very volatile at the baking temperature, full development is usually achieved by removal of the bisphenol A with isopropanol or aqueous base.

Preliminary lithographic experiments with this and similar activated polycarbonates, confirm that they possess enhanced sensitivity and afford good resolution. A typical example is shown in Figure 1. We have now extended this design concept to a variety of tertiary, allylic, and benzylic polycarbonates as well as to the corresponding polyethers and polyesters. These all share the same ability to undergo facile acid-catalyzed thermolysis, with formation of volatile low molecular weight fragments.

All of the previously described approaches involve chemical amplification which result in side-chain scission, or main chain cleavage with subsequent depolymerization. However, it is also possible to design materials which afford chemical amplification using processes in which chemical bonds, or entire polymer chains are created. We have reported an imaging system based on photochemical cross-linking of epoxy resins through ring opening polymerization initiated by a photochemically generated acid.[10] Perhaps the best illustration of this concept, however, is our development of photoinitiated interfacial catonic polymerization[21] and its application to imaging with chemical amplification. In this process, imaging is accomplished by casting a film of an inert polymer such as poly(p-methoxystyrene) containing a small amount of a soluble cationic photoinitiator, such as triphenylsulfonium hexafluoroantimonate. Here, exposure to radiation is used to create a latent image, which consists of strong acid moieties, dispersed in the inert polymer matrix. Development of the image is accomplished by treating the sample with an appropriate monomer, either in the liquid or gas phase. The photogenerated acid causes polymerization to occur in the areas of the polymer film which had initially been exposed to radiation, resulting in the formation of a negative image of the mask in a chemically amplified process. The kinetics of film growth in this process are such that

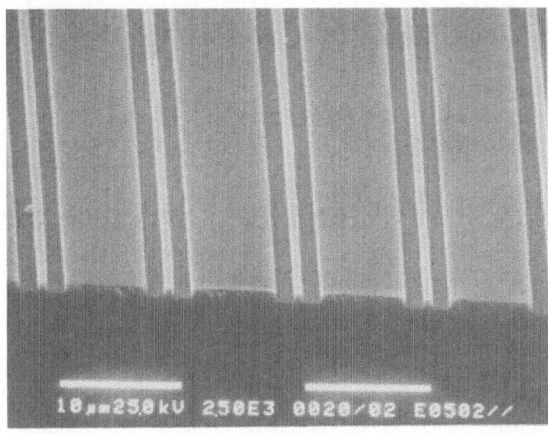

Figure 1. Deep-UV projection printed images in tertiary copolycarbonate. Smallest dimension is 1 micrometer.

good image quality can be obtained as a short induction period is followed by a rapid but limited film growth, with a stationery state being reached as the propagating chain ends are trapped within the growing film. Typically, this technique can be used to deposit films having a thickness of up to 1 μm. In practice, a full development to the substrate can be achieved by combining the use of a silcon-containing monomer for the film-growth step, with oxygen plasma development as shown in Figure 2.

As outlined in Figure 2, step #2 shows the creation of the latent image within the inert polymer coating, step #3 shows initial development through growth of a silicon-containing polymer on to the exposed areas of the coating, while step #4 shows the final development process. In this process, the deposited silicon-containing polymer acts as a plasma resistant film, protecting the exposed areas of the coating while the unexposed areas are etched away under the influence of an oxygen plasma. Figure 3a shows patterns obtained after interfacial cationic polymerization of p-trimethlsilylstyrene. Figure 3b shows the profile obtained after oxygen reactive ion etching of such a pattern. Excellent image quality can be achieved by this technique while only a low dose of radiation is required during the critical imaging step.[22]

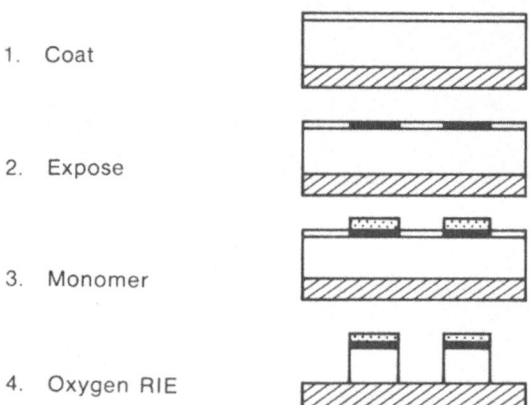

1. Coat

2. Expose

3. Monomer

4. Oxygen RIE

Figure 2. Photoinitiated interfacial cationic polymerization imaging process sequence.

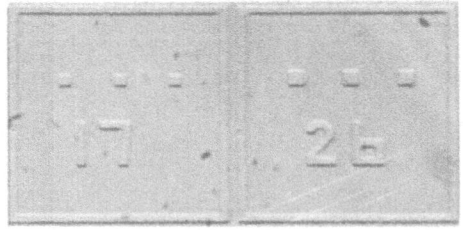

Figure 3a. Polymer deposited by photoinitiated cationic interfacial polymerization prior to oxygen etching.

CONCLUSION

Chemical amplification is a promising approach in the search for new systems which can be used in deep-UV imaging. The systems we have described show that the resist chemist can make a valuable contribution to microlithography through appropriate molecular design and the application of simple chemical concepts. Chemical amplification provides one potential solution to the problem of low flux in the current exposure equipment used in the deep-UV, as greatly enhanced resist sensitivities can be achieved with appropriate design. Although the ultimate resolution of such systems may be limited as excessive levels of amplification are reached, such concerns are not warranted as long as optical exposure is considered due to diffusion problems and chemical side-reactions which limit the distance over which amplification is effective. In practice, these distances are much shorter than the wavelength of the exposing light.

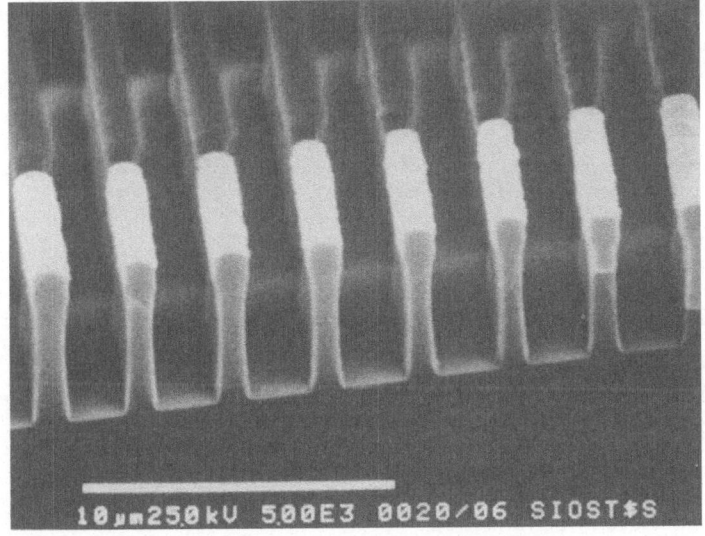

Figure 3b. High resolution, high aspect ratio images generated by photoinitiated cationic interfacial polymerization. Smallest dimension is 1 micrometer.

ACKNOWLEDGMENTS

Partial support of portions of this work done at the University of Ottawa by the Natural Science and Engineering Council of Canada and IBM under the GTA and SUR programs is gratefully acknowledged.

REFERENCES

1. L. F. Thompson, C. G. Willson, and M. J. Bowden, Eds., ACS Symposium Series 219 (1983).
2. (a) T. G. Tessier, J. M. J. Fréchet, C. G. Willson, and H. Ito, ACS Symposium Series 266, "Materials for Microlithography," L. F. Thompson, G. C. Willson, and J. M. J. Fréchet, Eds., American Chemical Society, Washington, D.C., p. 269 (1984); (b) C. G. Willson, H. Ito, J. M. J. Fréchet, T. G. Tessier, and F. M. Houlihan, *J. Electrochem. Soc.* 133:181 (1986).
3. J. Liukus, M. Hatzakis, J. Shaw, and J. Paraszczak, *J. Polym. Eng. Sci.* 23:1047 (1983).
4. H. S. Choong and F. J. Kahn, *J. Vac. Sci. Technol.* 19:1121 (1981).
5. S. Imamura, T. Tamamura, K. Harada, and S. Sugawara, *J. Appl. Polym. Sci.* 27:937 (1982).
6. Y. Harita, Y. Kamoshida, K. Tsutsumi, M. Koshiba, H. Toshimoto, and K. Harada, SPSE 22nd Fall Symposium, Washington D.C. (1982).
7. R. Tarascon, M. Hartney, and M. J. Bowden, ACS Symposium Series 266, *Materials for Microlithography,* L. F. Thompson, G. C. Wilson, and J. M. J. Fréchet, Eds., American Chemical Society, Washington, D.C., p. 361 (1984).
8. J. M. J. Fréchet, T. G. Tessier, C. G. Willson, and H. Ito, *Macromolecules* 18:317 (1985).
9. H. Ito and C. G. Willson, *Polym. Eng. Sci.* 23:1012 (1983).
10. H. Ito and C. G. Willson, ACS Symposium Series 242, "Polymers in Electronics," T. Davidson, ed., Amer. Chem. Soc., Washington D.C., p. 11 (1984).
11. H. Ito, C. G. Willson, and J. M. J. Fréchet, Digest of Technical papers of 1982 Symposium on VLSI Technology, Olso, Japan, p. 86.
12. H. Ito, C. S. Willson, and J. M. J. Fréchet, U.S. Patent 4, 491,628, assigned to IBM Corporation (1985).
13. J. V. Crivello, "UV Curing: Science an Technology," S. P. Pappas, Eds., Technology Marketing Corporation, Stamford, Connecticut, p. 23 (1978).
14. H. C. Beyerman and J. S. Bonteboe, *Proc. Chem. Soc.,* 249 (1961).
15. H. C. Beyerman and G. J. Herszwolf, *J. Chem. Soc.,* 755 (1963).
16. G. A. Fletcher and J. H. Jones, *Intl. J. Dept. Protein Res.* 4:347-371 (1972).
17. J. M. J. Fréchet, Abstracts of the Australian USH Workshop on Radiation Effects on Polymeric Materials, NOOSA, Australia, p. 16 (1987).
18. E. Reichmanis and J. H. O'Donnell, Radiation Effects on Polymeric Materials, Amer. Chem. Soc., Washington D.C. (in press).
19. F. M. Houlihan, F. Bouchard, J. M. J. Fréchet, and C. G. Willson, *Macromolecules* 19:13 (1986).
20. J. M. J. Fréchet, E. Eichler, M. Stancinlessu, T. Iizowa, F. Bouchard, F. M. Houlihan and C. G. Willson, "Acid Catalyzed Thermolytic Depolymerization of Polycarbonates: A New Approach to Dry-Development Resist Materials," Chap. 11, in "Polymers for High Technology," ACS Symposium Series 346, M. J. Bowden and S. R. Turner, Eds., Amer. Chem. Soc., Washington D.C. (1987)
21. A. Hult, S. A. MacDonald, and C. G. Willson, *Macromolecules* 18:1804 (1985).
22. S. A. MacDonald, A. Hult, R. D. Allen, and C. G. Willson, Proc. XIth Int. Conf. in Org. Coatings, Sci. and Technol. Proceedings 11, p. 203 (1985).

CLOSING COMMENTS

Steve Corley

Shell Development Company — WRC

P.O. Box 1380, Houston, TX 77251-1380

Thanks, Professor Martell. It seems that I've been nominated to be the one who gets to talk at the tail end of this Symposium, so I'd better make it quick or else you'll all walk out -- and I don't blame you. I hope you've enjoyed the past 2-1/2 days here in Aggieland. I certainly have appreciated all of the excellent and very informative talks you've given. I hope that while you've been here you've learned one of the traditions of this area, which is the Aggie joke. I think I will add one to the inventory of Aggie jokes. As you know, if you looked at some of the descriptive material in your registration packet, the Aggie mascot is allegedly a little collie called "Reveille." But that statement is actually not true. Aggies originally got their name from the *real* Aggie mascot, who is called "Aggie." And why is the mascot called "Aggie?" "Aggie" is short for *Agkistrodon piscivorus,* which is the zoological classification of the cottonmouth, or water moccasin -- this being cottonmouth country. But, one might ask, what about northern schools? Some of their teams are called "Aggies" also and there are no cottonmouths in the north. Well, the answer to that question is simple. The teams from northern states which are called "Aggies" use as their mascot *Agkistrodon contortrix,* the copperhead.

Yesterday evening Professor Vogl gave in his banquet lecture a rather extensive presentation which summarized what constitutes a functional polymer and what functional polymers do. I would like to add a few personal comments on that, which I believe will help to put some things in historical perspective. The development of human civilization over the past few millennia closely parallels the growth in man's use of metals. This began with the use of native copper and gold nuggets which were pounded into objects valued for their mechanical properties. Man then learned to smelt copper and tin ores and to make bronze, and subsequently to purify the much stronger iron. These metals were used, first in tools, and then in structures, with increasingly better mechanical properties. This progression eventually made possible the development of large steel structures, the steel buildings which are so commonplace in today's society.

Over the last 200 years, however, metals have also been increasingly used for many purposes that do not rely on their mechanical properties. One of the earliest of these uses was as metallic or metal-based pigments. Paralleling the growth of the large-scale use of electricity in the past century has been the use of metals for their electrical properties, to form our modern net of electrical conductors covering much of the world. The use of metal halides in photography, for example, has nothing to do with the metals' mechanical properties. We had a session on polymer blends earlier during this symposium. Long before man made his first polymer blend, however, he was using metals as alloying agents to make metal blends with better properties than the original metals. Of course, the use of metals as catalysts has also grown to enormous importance.

Ceramics have followed a similar course, beginning with the ancient Mesopotamians who used bricks and made brick buildings. Cement, concrete, and to some extent glass have similarly been used largely for their mechanical properties. Paralleling the broadening of the uses of metals, however, ceramics have also over the last two centuries developed many uses not related to their mechanical properties. These include ceramic insulators, piezoelectric ceramics, solar cells, zeolite catalysts, and most recently the superconducting ceramics.

I think that the presentations given in this symposium have clearly shown that polymers are simply following the same natural upward course as metals and ceramics. Man very early began to use natural polymers, such as cellulose fibers, animal fibers, and later rubber. Then he began to modify these natural materials somewhat (e.g., vulcanized rubber, cellulose nitrate). Around the turn of the century came the development of phenolic resins, one of the first classes of purely synthetic polymers. Like the natural materials, they were used for their mechanical properties. Then followed, as Professor Vogl covered yesterday evening, the rapid development during the early and middle part of this century of the large volume thermoplastic and thermosetting materials -- likewise used basically as structural components, for their mechanical properties.

Then, however, came the growth of functional uses of polymers, as has been so well covered by all the speakers in this Symposium -- paralleling the earlier uses of metals and ceramics for these purposes. Examples of these uses were given by the main topic headings of this Symposium -- the uses of polymers as blending agents for other polymers, the biomedical uses, uses in membranes, uses as electrical conductors and elements of electronic circuits, uses as catalysts and catalyst supports, and the use of polymers for their photoresponsive properties. All of these areas speak for themselves and show that polymers are following the same course earlier followed by metals and ceramics for uses in human civilization. We can see that with these increasing functional uses, polymers are truly coming into their own as basic materials of man and his culture, as full peers of the older materials (metals and ceramics). And the growth curve of polymers, polymer value-added uses, and polymer applications continues ever upward.

And on this note, speaking for Texas A&M University and for the IUCCP, I would like to conclude this Symposium. Thank you all -- I hope you enjoyed this Symposium as much as I did -- and have a safe trip home.

ACKNOWLEDGMENTS

I would certainly like to thank a lot of people who have contributed so much
to the success of this symposium. This includes the speakers, all of you, in
the sessions, and particularly Professor Vogl, who gave the banquet address.
All of the session chairmen, who did their best to keep the program on track
and moving smoothly. This also includes the companies who sponsored the IUCCP
(and whose sponsorship of the IUCCP made the symposium possible), Texas A&M
University itself, the Chemistry Department and other sponsoring departments
of the University, and particularly Professor Martell, who went out of his way
in his work in organizing this symposium. It also includes those who organized
the banquet and participated in the banquet such as the hotel employees, the
questioners and all of you active listeners who did so much to make each
presentation most effective in presenting information. And in particular I
would like to thank a number of the graduate students who did so much. I
would like to thank Cynthia McBride, the secretary, who took a particularly
active role in organizing this symposium and putting together the things that
needed to be done, the graduate student workers on this list and particularly
those who drove the bus which took us back and forth between the hotel and the
symposium sessions. These include Mike Bothwell, Chris Campbell, Billy
Curtis, Mark Espenscheid, Leif Ergens, Susan Michelbaugh, and Irene Oshiro.
(I'm sorry if I omitted anyone's name. If I did, I certainly did it
inadvertently.)

POLYETHYLENE-BOUND TIN HALIDE AS A CATALYST FOR REDUCTIONS

OF ALKYL HALIDES USING NaBH$_4$

Samuel A. Walker and David E. Bergbreiter

Department of Chemistry, Texas A&M University

College Station, Texas 77843

Abstract

The synthesis, characterization and use of polyethylene-bound dibutyltin chloride as a catalyst for the reduction of alkyl halides using a suspension of sodium borohydride in toluene and a crown ether as a phase transfer catalyst is described. This tin-containing soluble macromolecule was synthesized by anionic oligomerization of ethylene followed by electrophilic substitution of the resulting "living" oligomer with dibutyltin dichloride. The resulting polyethylene-bound tin reagent was characterized by NMR and IR spectroscopy. It was used as a cocatalyst (5-10%) along with 10-20% of a phase transfer catalyst (usually a crown ether) in reductions of alkyl halides using a suspension of sodium borohydride in hot toluene at 110 °C. Preliminary kinetic studies imply that the reaction rate is dependent on the rate of transport of sodium borohyride into the toluene phase by the crown cocatalyst. Yields of reduced products are typically > 90 %.

FUNCTIONALIZED POLY(ALKYL/ARYLPHOSPHAZENES)

Patty Wisian-Neilson,* Mary Alice Schaefer, and Randal R. Ford

Department of Chemistry, Southern Methodist University,

Dallas, Texas 75275

Abstract

Poly(methylphenylphosphazene) can be derivatized through either deprotonation/substitution reactions at the methyl group or electrophilic aromatic substitution of the phenyl group. These reactions have been used to prepare a variety of new polyphosphazenes with all functional groups attached to the polymer backbone by direct P-C bonds. The synthesis and characterization of several of these new

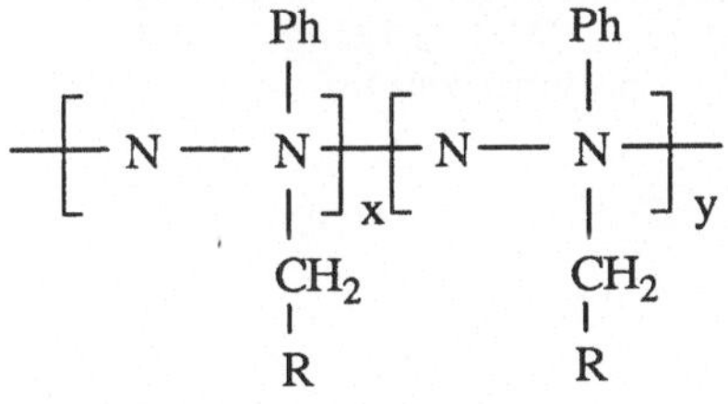

$R = R'Me_2Si$ [R' = H, Me, vinyl, $(CH_2)_2CN$], Ph_2P, alkyl, COOH, $[CH_2CH(Ph)]_nH$, and $R'R''C(OH)$ [R', R'' = H, Me, Ph, ferrocene, and thiophene]

copolymers will be discussed.

VARIABLE-TEMPERATURE NMR STUDIES OF POLYPHOSPHAZENES

Richard C. Crosby, James F. Haw, Shawn J. Maynard, and Ronnie Reese

Department of Chemistry, Texas A&M University

College Station, Texas 77843

Abstract

Polyphosphazenes are a diverse class of inorganic polymers based on chains of alternating phosphorus and nitrogen atoms. These polymers have unique physical and chemical properties which make them of interest to chemists and material scientists. We report variable-temperature (VT), multinuclear, NMR results which address several important problems in polyphosphazene chemistry. P-31 and C-13 VT magic-angle spinning (MAS) NMR are being used to study morphology, cross-linking, and thermal degradation of these polymers. Solid-state H-1 NMR is being used to examine the effects of water on the melt polymerization of the hexachlorocyclotriphosphazene.

MATHEMATICAL MODELING OF ELECTRONICALLY

CONDUCTIVE POLYPYRROLE

Taewhan Yeu, Jose Carbajal, and Ralph E. White*

Department of Chemical Engineering, Texas A&M University

College Station, Texas 77843-3122

Abstract

A mathematical model for predicting cyclic voltammograms on a rotating disk electrode (RDE) with polypyrrole is presented. The model is used to predict the spatial and time dependence of concentration, overpotential, and stored charge profiles within the system, and to clarify the electrochemical reaction mechanisms and transport phenomena within the polypyrrole film. The model includes both faradaic and capacitive charge components into total current expression. Also, a mathematical model for the lithium/lithium perchlorate - propylene carbonate/polypyrrole (Li/LiClO$_4$-PC/PPy) secondary battery cell is expanded from the RDE model. This allows for a better understanding of the dynamic behavior and provides guidance toward design of new secondary batteries utilizing electronically conductive polypyrrole cathodes. The model includes the capabilities of handling charge and discharge behavior. The model is used to study the effects of various design parameters on the performance of the cell.

SYNTHESIS OF GOLD CONTAINING FUNCTIONALIZED POLYMERS

Md. Nazrul I. Khan, John P. Fackler, and J.C. Wang

Laboratory of Molecular Structure and Bonding, Department of Chemistry

Texas A&M University, College Station, Texas 77843-3122

Abstract

Modification of structure-property relationships of functionalized polymers (e.g., ethylene acrylic acid copolymer, EAA) by the introduction of dispersed heavy metal atoms or small clusters can lead to materials with interesting new reflectivity and energy scattering properties. To date, research in these areas have not been well explored. Gold and heavy elements close to it in the periodic table are particularly interesting as modifiers due to their special reflectivity properties for various type of high energy radiation, including accelerated particles. In recent years the introduction of gold as $HAuCl_4$ into polyimide films has been shown to modify the thermal properties of this polymer.

In the present studies, two functionalized polymers (supplied by Dow Chemicals) such as PRIMACOR 1430 (containing 9.5% acrylic acid) and PRIMACOR 5980 (containing 20% acrylic acid) have been used for the synthesis of polymer/gold dispersion (10 to 12 wt % Au) and polymer/gold ionomer (17 to 20 wt % Au) respectively. Organometallic gold compounds such as $Ph_3PAuC_6F_5$ with the P1430 polymer and Ph_3AuNO_3 with the P5980 polymer for ionomer have been used. These polymers P1430 (dispersion) and P5980 (ionomer) were dissolved in xylene and xylene/THF (3:1) mixture at 80 -100° C under normal pressure. The compounds $PH_3PAuC_6F_5$ and Ph_3PAuNO_3 were added to the homogeneously clear solution of P1430 and P5980 respectively at higher temperature. In both cases (under an inert atmosphere) the solution mixtures were refluxed for 2 h and then the solvent was evaporated under reduced pressure (vacuum pump) to yield a colorless solid with the P1430 polymer and slightly colored (light purple) solid with the P5980 polymer respectively. Neutron activation analysis shows these polymers contain 10.8 - 11.2 wt % Au dispersed in the case of P1430 and 12-14% in the case of P5980. A scanning electron micrograph (SEM) was used to determine the distribution pattern of Au in the polymer.

THE ELECTRODE/ELECTROLYTE INTERPHASE OF ELECTRONICALLY CONDUCTING POLYPYRROLE

David L. Miller and John O'M. Bockris

Department of Chemistry, Texas A&M University

College Station, Texas 77843

Abstract

As part of an overall study of the electrode/electrolyte interphase of the electronically conducting polymer, polypyrrole, the surface structure and electronic properties have been investigated.

Scanning electron microscopy was performed on polypyrrole to determine what the surface structure looked like on a magnified view. The surface is rough and has a cauliflower-like appearance. However, the structure is fairly uniform.

The apparent surface area and pore size distribution was measured for polypyrrole films. The method used involved adsorption of solute molecules from aqueous solution onto the polymer surface. Different size adsorbates were utilized to provide a range of apparent surface areas as a function of polymer pore size. The results indicate that for adsorbates capable of entering pores of radii in the range of 4.0 to 6.5A, a roughness factor, or ratio of apparent to geometric surface area, of 140 to 150 is found. An adsorbate capable of entering into pores of 3.2 A radii generated a roughness factor of 600. This indicates a large number of pores in polypyrrole having radii in the 3.2 to 4.0 A range.

Carrier densities and carrier mobilities were obtained for polypyrrole via the Hall effect technique. The apparatus was also used to measure conductivities. The analysis was performed for polypyrrole containing three different counterions: p-toluenesulfonate, perchlorate, and tetrafluoroborate. The carrier density was found to vary by one order of magnitude, from 10^{19} to 10^{20} cm-3. The carrier mobility, on the other hand, remained constant with respect to counterion, at 1 $cm^2V^{-1}sec^{-1}$. In order to ensure that any solvent from preparation of the polymer was not affecting the results, one polymer, polypyrrole-p-toluenesulfonate, was subjected to vacuum dehydration for varying periods of time, followed by re-analysis. It was found that the carrier density changes very little with dehydration. There was no affect on the carrier mobility.

ORGANIC POLYMERS IN ENZYME IMMOBILIZATION

Gordon L. Smith and C.-H. Wong

Department of Chemistry, Texas A&M University

College Station, Texas 77843

Abstract

Structural stability is a critical factor in the practical application of enzymes as catalysts. Evidence based on structural analysis has shown that thermophilic enzymes usually differ from their mesophilic counterparts by only insignificant alteration of their primary structure, and the three dimensional structure of such enzymes are largely similar. The observed difference in operational temperature corresponds to an increase in stability by 5-7 kcal/mol. A free energy change of this order could be derived from one additional salt bridge or several additional hydrogen bonds inside the protein molecule. Perception of these differences and their structural determinants would allow manipulation of proteins to produce stability enhancement. Many methods of enzyme stabilization have been developed but none of them is generally applicable. These include covalent bonding to activated supports, physical adsorption to support matrixes, electrostatic entrapment, and suspension in semipermeable membranes. Another method used to stabilize enzymes is to introduce additional internal binding forces via site-specific mutagenesis. Procedures are described here for the preparation of water soluble immobilized enzymes of interest as catalysts in organic synthesis. Multipoint attachment between the enzyme and the water-soluble polymer poly(acrylamide-co-N-acrylamide- co-N-acryloxysuccinimide), PAN,was used to stabilize the enzyme in solution. This approach takes advantage of the flexibility of the polymer to react in a manner which keeps disruption of the preferred enzyme conformation to a minimum. Application of this technique was demonstrated with α-chymotrypsin, trypsin, FDP-aldolase, pig liver esterase, horse liver alcohol dehydrogenase, and lactate dehydrogenase. Enhancements in enzyme stability of greater than 100 times were observed in some cases. Model studies indicated that the PAN polymer reacts selectively with primary amines. These results can be related to the number and type of nucleophile containing amino acids on the surface of the enzyme and allow for a better determination of the optional reaction conditions for increased enzymatic stability. A new copolymer of styrene and acryloxysuccinimide was also prepared and used to immobilize several enzymes. Unlike the water-soluble PAN, this copolymer is insoluble in water but soluble in many organic solvents. The crosslinked enzymes were tested at 25°C in dioxane and water mixtures, and other organic solvents. In all cases, the immobilized enzyme was found to retain much greater stability than the corresponding non-treated enzyme.

POLYETHYLENE FUNCTIONALIZATION USING METAL CARBOXYLATES

Jerry L. Poteat and D.E. Bergbreiter

Department of Chemistry, Texas A&M University

College Station, Texas 77843

Abstract

The synthesis of copper(II) functionalized polyethylene powders by entrapment functionalization is described along with preliminary studies of the reactivity of copper(II) ions located at the polymer-solution interface of such functionalized powders. Ethylene oligomers terminated with copper(II) carboxylates were prepared by anionic oligomerization of ethylene using n-butyllithium-TMEDA as an initiator followed by quenching with carbon dioxide. The carboxylic acid resulting from protonation of this product was allowed to react with copper(II) naphthoxyglycolate to form a copper carboxylate of polyethylene. Dissolution of this copper containing oligomer along with virgin polyethylene followed by cooling formed the entrapment functionalized powder. Reactions of the copper in this entrapment functionalized powder with complexing agents including diphenylthiocarbazone and TRIAP were studied by UV-visible and fluorescence spectroscopy, respectively. Comparison of the amount of copper(II) extracted within 12 h with either of these reagents with the amount of total copper(II) estimated to be present based on ^{1}H NMR spectroscopy (of a methyl ester of the starting oligomer) or based on ICP analysis showed that a substantial amount of the entrapped copper(II) was at the polymer solution interface. As expected, ESCA too showed copper was present at the surface of these functionalized films.

ACID-BASE CHEMISTRY AT POLYMER-SOLUTION INTERFACES

M.D. Hein, K.J. Huang and D.E. Bergbreiter

Department of Chemistry, Texas A&M University

College Station, Texas 77843

Abstract

Surface functionalized polyethylene films have been prepared by entrapping ethylene oligomers terminated with N,N-dimethyl *para*-aminobenzeneazobenzoic acid (*p*-methyl red). These surface functionalized films were exposed to sulfonic acids in various solvents to evaluate the accessibility, reactivity and mobility of functional groups at the polymer surface. Utilizing methane sulfonic acid in different solvents, the appearance of the protonated form of the dye in the film was monitored by UV-visible spectroscopy. Relative percentages of protonation were then calculated. The accessibility of functional groups at the polymer surface toward methane sulfonic acid in solution was shown to be a function of the dielectric constant of the solvent used. Solvents with low dielectric constants (e.g. CHCl$_3$) which are more compatible with polyethylene allow the sulfonic acid greater access to the dye groups. Solvents with higher dielectric constants (e.g. acetone) protonate smaller amounts of the entrapped dye. The reactivity of the entrapped dye was also influenced by being at a polyethylene-solution interface. While a low molecular weight analog of the entrapped dye, the octyl ester of *p*-methyl red, was stable as its protonated from in solvents leveling effect was observed. The entrapped protonated dye was more acidic than protonated solvent and thus it transferred its proton to the solvent. This was observed to be a rapid process. This proton transfer was verified by using soluble dyes as proton acceptors. Finally the mobility of these azo dye molecules could be studied. The rate of cis-trans isomerization of these bound azo dyes was shown to be only slightly dependent on the solvent in which the polyethylene film was suspended.